AF431624

Introduction to Elementary Mathematics

— Grades 2 and 3 —

Part of the
Mastering Essential Math Skills Series

**For access to our FREE
online video tutorials, go to
www.mathessentials.net**

Richard W. Fisher

Introduction to Elementary Mathematics: Grades 2 and 3

Copyright © 2023, Richard W. Fisher
All rights reserved. No part of this publication may be reproduced, stored in a retrieval system or
transmitted in any form or by any means, electronic, mechanical, photocopying, recording, or otherwise,
without the written permission of Richard W. Fisher.

Manufactured in the United States of America

Softcover ISBN: 979-8-9888207-1-0

1st printing 2023

Math Essentials
4177 Texas Elm
San Antonio, TX 78230

www.mathessentials.net
math.essentials@verizon.net

Some Tips for the Student

■ Each lesson in this book includes a video tutorial. For the best possible learning experience the student should go through the video tutorial first, before going to the lesson in the book.

■ Don't just watch the video. Work right along with the instructor. This will prepare the student and make the lesson in the book much easier to understand.

■ Also, invest in a graph paper notebook. This will help the student to be organized and also to complete their work neatly. Being neat will help to avoid careless errors.

Notes to the Teacher or Parent

What sets *Mastering Essential Math Skills* apart from other books is its approach. It is not just a math book, but a *system* of teaching math. Each daily lesson contains five key parts: two speed drills, review exercises, teacher tips (Helpful Hints), a section containing new material, and a daily word problem. Teachers have flexibility in introducing new topics, but the book provides them with the necessary structure and guidance. The teacher can rest assured that essential math skills are being systematically learned.

With so many concepts and topics in the math curriculum, some of these essential skills are easily overlooked. This easy-to-follow math program requires only twenty minutes of instruction per day. Each lesson is concise, and self-contained. The daily exercise help students to not only master math skills, but also to maintain and reinforce those skills through consistent review—something that is missing in most math programs. Skills learned in this book apply to all areas of the math curriculum, and consistent review is built into each daily lesson. Teachers and parents will also be pleased to note that the lessons are quite easy to correct.

The book is divided into eight chapters, which cover whole numbers, fractions, decimals, percentages, integers, geometry, charts and graphs, and problem solving.

Mastering Essential Math Skills is based on a system of teaching that was developed by a math instructor over a twenty-year period. This system has produced dramatic results for students. The program quickly motivates students and creates confidence and excitement that leads naturally to success.

Please read the following "How to Use This Book" section and let this program help you to produce dramatic results with your children or math students.

How to Use this Book

Mastering Essential Math Skills is best used on a daily basis. The first lesson should be carefully gone over with the students to introduce them to the program and familiarize them with the format. A typical lesson has been broken down on the following pages into steps to suggest how it can best be taught. It is hoped that the program will help your students to develop an enthusiasm and passion for math that will stay with them throughout their education.

As you go through these lessons every day you will soon begin to see growth in the students' confidence, enthusiasm, and skill level. The students will maintain their mastery through the daily review.

In school, the book is best used during the first part of the math period. The structure and format seem to naturally condition the students to "think nothing but math" from the moment class begins. The students are ready to "jump into the lessons" without any prompting or motivating needed from the teacher. This makes for a very smooth and orderly start each day.

Also, once you have finished the daily lesson, there will still plenty of time to explain related topics, or work on new topics in the basic test or through other sources.

3

It is illegal to photocopy this page. Copyright © 2023, Richard W. Fisher

Step 1

Students open their books to the appropriate lesson and begin together. Have students first go to the review exercises, working each problem and showing all their work. If students finish early, they are to check their work in the review section.

Step 2

When you feel that enough time has been spent on the review exercises (usually two to three minutes), call out "Time." The next step is to go to the speed drills. A good signal is to say, "Get ready to add." The students go to the addition drill and wait for the next signal. Then say, "The number to add is '( )'." At this stage the students place the given number inside the addition circle and, as quickly as possible, write all the sums in the appropriate space outside the perimeter of the circle. As students complete the drill, have them drop their pencils and stand or signal in some appropriate way. When enough time has been given, say, "Time." Students then correct the drill as the answers are read aloud by the teacher or a student. The same process is used for the multiplication drill. It is amazing how motivating these speed drills can become in helping students to master their addition and multiplication facts.

Step 3

After the speed drills, work through the review problems with the class. Work the problems on the board or overhead and go through them step-by-step with the students, drawing responses and asking questions as you go. Allow the students to check their own work in this section. This section provides consistent review and reinforcement of topics that the class learns.

Step 4

After going through the review exercises, give a short introduction of the new material. This is where the teacher's unique style and skills come into play. Appropriate concepts, vocabulary, and skills can be introduced on the blackboard or overhead. This should require only a few minutes.

4

It is illegal to photocopy this page. Copyright © 2023, Richard W. Fisher

It is illegal to photocopy this page. Copyright © 2023, Richard W. Fisher

Step 5

After a brief introduction of the new material, go over the "Helpful Hints" section with the class. Be sure to point out that it is often helpful to come back to this section as the students work independently. This section often has examples that are very helpful to the student.

Step 6

After going through the "Helpful Hints" section, do the two sample problems. It is highly important to work through these two problems with the class. The students can model the techniques that are discussed and demonstrated by the teacher. Go through the steps on the board or overhead, and the students can write them directly into their books. Working these sample problems together with the class can prevent a lot of unnecessary frustration on the part of the students. In essence, in working them together, each student has successfully completed the first two problems of the lesson. This can assist in developing confidence as a routine part of each daily lesson.

Step 7

Next, allow the students to complete the daily exercises and the word problem of the day. Make it a point to circulate and offer individual help. If it is necessary, work another example or two on the board with the entire class. Also, reading the word problem of the day together with the class before they work it independently may be very beneficial.

Step 8

Last, collect the books, correct them and return them the next day. It may sometimes be appropriate to correct them with the students.

Contents

Whole Numbers

Addition 9
Addition with Large Numbers 10
Subtraction 11
Subtraction with Zeroes 12
Reviewing Addition and Subtraction 13
Multiplying by 1-Digit Factors 14
Multiplying by 2-Digit Factors 15
Reviewing Multiplication of Whole Numbers 16
Division by 1-Digit Divisors 17
Dividing Hundreds by 1-Digit Divisors 18
Reviewing Division by 1-Digit Divisors 19
Reviewing All Whole Number Operations 20

Fractions

Identifying Fractions 21
Expressing Fractions in their Simplest Forms 22
Changing Improper Fractions to Mixed Numerals 23
Adding with Like Denominators 24
Adding Mixed Numerals with Like Denominators 25
Subtracting with Like Denominators 26
Subtracting Mixed Numerals or Fractions from Whole Numbers 27
Subtracting Mixed Numerals with Like Denominators 28
Reviewing Adding and Subtracting Mixed Numerals 29
Finding the Least Common Denominator 30
Adding and Subtracting with Unlike Denominators 31
Adding Mixed Numerals with Unlike Denominators 32
Reviewing Addition and Subtraction of Fractions and Mixed Numerals 33
Multiplying Common Fractions 34
Multiplying Common Fractions / Eliminating Common Factors 35
Multiplying Whole Numbers and Fractions 36
Multiplying Mixed Numerals 37
Reciprocals 38
Dividing Fractions and Mixed Numerals 39
Reviewing All Fraction Operations 40

Decimals

Reading Decimals 41
Changing Words to Decimals 42
Changing Fractions and Mixed Numerals to Decimals 43
Changing Decimals to Fractions and Mixed Numerals......................... 44
Comparing Decimals 45
Adding Decimals 46
Subtracting Decimals 47
Reviewing Addition and Subtraction of Decimals 48
Multiplying by Whole Numbers 49
Multiplying Decimals 50
Reviewing Multiplication of Decimals 51
Dividing Decimals by Whole Numbers 52
Changing Fractions to Decimals 53
Reviewing Dividing Decimals 54
Reviewing All Decimal Operations 55

Ratios, Proportions, Percent

Ratios 56
Changing Tenths and Hundredths to Percents 57
Changing Decimals to Percents 58
Changing Percents to Decimals and Fractions 59
Finding the Percent of a Number 60
Finding the Percent of a Number in a Word Problem 61
Changing Fractions to Percents 62
Finding the Percent 63
Finding the Percent in a Word Problem 64
Reviewing Percent Problems 65
Reviewing Percent Word Problems 66
Reviewing Ratios, Proportions, and Percents 67

It is illegal to photocopy this page. Copyright © 2023, Richard W. Fisher

Geometry

Points, Lines, Planes 68
Parallel, Intersecting Lines, Angles 69
Types of Angles 70
Measuring Angles 71
Measuring Angles with a Protractor 72
Polygons 73
Classification of Triangles 74
Perimeter of Polygons 75
Circles 76
Circumference of Circles 77
Areas of Squares and Rectangles 78
Reviewing Perimeter, Circumference, and Area 79
Solid Figures 80
Reviewing Geometry 81

Number Theory and Algebra

Factors of a Number 82
Greatest Common Factor 83
Multiples of a Number 84
Least Common Multiples 85
Algebraic Equations with Addition 86
Algebraic Equations with Subtraction 87
Algebraic Equations with Multiplication 88
Algebraic Equations with Division 89
Reviewing Algebraic Equations 90
Reviewing Number Theory and Algebra 91

Integers

Addition 92
More Addition 93
Subtraction 94
Reviewing Addition and Subtraction 95
Multiplication 96
Division 97
Reviewing All Integer Operations 98

Charts and Graphs

Bar 99-100
Line 101-102
Circle 103-104
Picture 105-106
Graphing Points on a Number Line 107-108
Graphing Ordered Pairs 109-110
Reviewing Charts and Graphs 111
Reviewing Graphing on a Number Line and Ordered Pairs 112

Probability and Statistics

Probability 113
More Probability 114
Range and Mode 115
Mean and Median 116
Reviewing Range, Mode, Mean and Median 117
Reviewing Probability and Statistics 118

Word Problems

1-Step Word Problems 119
1-Step Fractions 120
1-Step Decimals 121
1-Step Word Problem Review 122

Review Section

Final Reviews 123-132
Answer Key 133-150
Resource Center 151-160

8

It is illegal to photocopy this page. Copyright © 2023, Richard W. Fisher

Review Exercises	Speed Drills

1.   42
   + 31

2.   713
   + 24

3. $6 + 7 + 4 =$

4.   5
    2
 + 3

1. Line up the numbers on the right side.
2. Add the ones first.
3. Remember to regroup when necessary.
4. "Sum" means to add.

HELPFUL HINTS

S.   23
  + 32

S.   423
  + 345

1.   43
  + 54

2.   62
  + 33

3.   42
  + 36

4.   324
  + 83

5.   426
  + 314

6.   453
  + 232

7. $33 + 24 + 16 =$

8. $34 + 216 =$

9. Find the sum of $223 + 234$

10. Find the sum of 12 and 18

1	
2	
3	
4	
5	
6	
7	
8	
9	
10	
Score	

Problem Solving There are 34 students in Mr. Smith's class, and 33 in Ms. Garcia's class. What is the total number of students?

It is illegal to photocopy this page. Copyright © 2023, Richard W. Fisher

Speed Drills

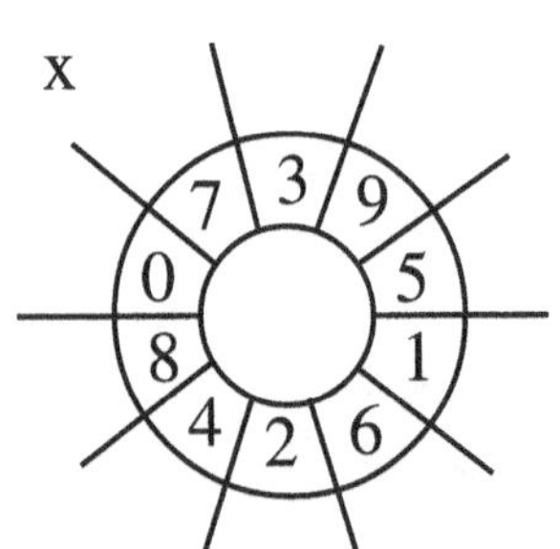

HELPFUL HINTS

Review Exercises

1.
$$
\begin{array}{r}
34 \\
12 \\
+\ 26 \\
\hline
\end{array}
$$

2.
$$
\begin{array}{r}
315 \\
+\ 24 \\
\hline
\end{array}
$$

3. $42 + 116 =$

4. Find the sum of 18 and 24

When writing large numbers, place commas every three numerals, starting from the right side. This makes them easier to read.

Example: 5 million, 234 thousand, 216

5,234,216

Score	
	1
	2
	3
	4
	5
	6
	7
	8
	9
	10

S.
$$
\begin{array}{r}
342 \\
+\ 237 \\
\hline
\end{array}
$$
S.
$$
\begin{array}{r}
3,762 \\
+\ 514 \\
\hline
\end{array}
$$
1.
$$
\begin{array}{r}
2,437 \\
+\ 2,564 \\
\hline
\end{array}
$$
2.
$$
\begin{array}{r}
536 \\
+\ 16 \\
\hline
\end{array}
$$

3.
$$
\begin{array}{r}
5,232 \\
+\ 1,423 \\
\hline
\end{array}
$$
4.
$$
\begin{array}{r}
2,736 \\
+\ 5,521 \\
\hline
\end{array}
$$
5.
$$
\begin{array}{r}
7,213 \\
+\ 2,314 \\
\hline
\end{array}
$$
6.
$$
\begin{array}{r}
2,134 \\
+\ 3,213 \\
\hline
\end{array}
$$

7. Find the sum of 2,132 and 223

8. $3,512 + 4,213 =$

9. $462 + 374 =$

10. Find the sum of 1,702 and 337

10 One town has a population of 523 and another has a population of 412. What is the total population of the two towns?

Problem Solving

It is illegal to photocopy this page. Copyright © 2023, Richard W. Fisher

Review Exercises	Speed Drills

1. $\begin{array}{r} 23 \\ + 42 \\ \hline \end{array}$ 2. $\begin{array}{r} 2{,}314 \\ + 3{,}214 \\ \hline \end{array}$

3. Find the sum of 372 and 314

4. $16 + 172 =$

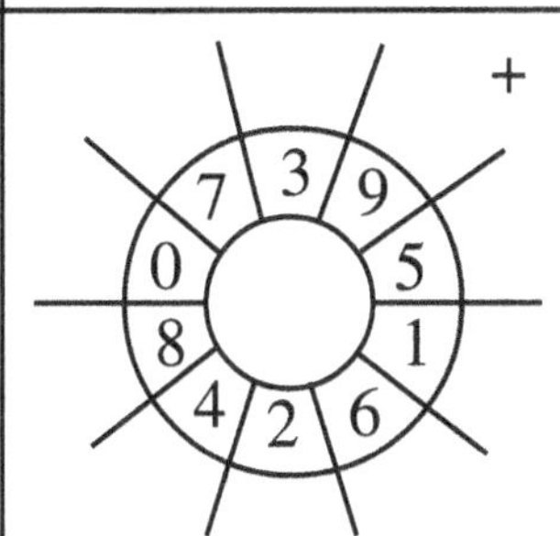

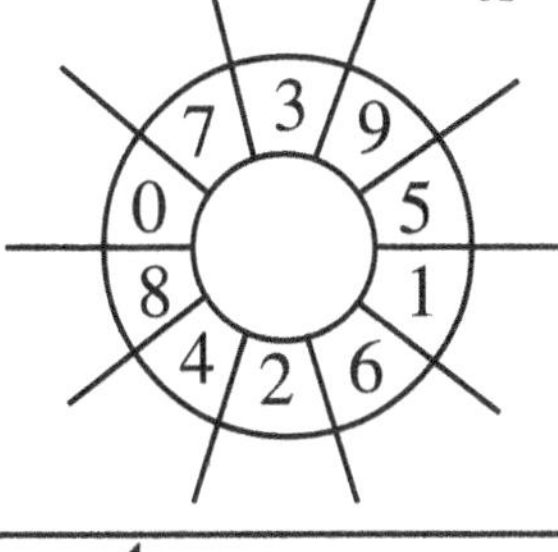

Examples

1. **Line up the numbers on the right side.**
2. **Subtract the ones first.**
3. **Remember to regroup when necessary.**
4. **It may be necessary to regroup more than once.**
5. **"Find the difference" means to subtract.**
6. **"Show how much more" means to subtract.**

$\begin{array}{r} {}^{8}\;{}^{1} \\ 7\,\cancel{9}\,3 \\ -\;7\,5 \\ \hline 7\,1\,8 \end{array}$ $\begin{array}{r} {}^{5}\;{}^{11}\;{}^{1} \\ \cancel{8}\,\cancel{2}\,3 \\ -\,2\,5\,4 \\ \hline 3\,6\,9 \end{array}$

HELPFUL HINTS

S. $\begin{array}{r} 356 \\ -\;123 \\ \hline \end{array}$ S. $\begin{array}{r} 352 \\ -\;171 \\ \hline \end{array}$ 1. $\begin{array}{r} 37 \\ -\;22 \\ \hline \end{array}$ 2. $\begin{array}{r} 64 \\ -\;22 \\ \hline \end{array}$

3. $\begin{array}{r} 655 \\ -\;232 \\ \hline \end{array}$ 4. $\begin{array}{r} 765 \\ -\;82 \\ \hline \end{array}$ 5. $\begin{array}{r} 322 \\ -\;151 \\ \hline \end{array}$ 6. $\begin{array}{r} 753 \\ -\;25 \\ \hline \end{array}$

7. Find the difference between 134 and 22.

8. Subtract 336 from 847.

9. $877 - 625 =$

10. 75 is how much more than 23?

1	
2	
3	
4	
5	
6	
7	
8	
9	
10	
Score	

Problem Solving 76 kids walk to school and 23 take the bus. How many more walk to school than take the bus?

11

It is illegal to photocopy this page. Copyright © 2023, Richard W. Fisher

Speed Drills	Review Exercises

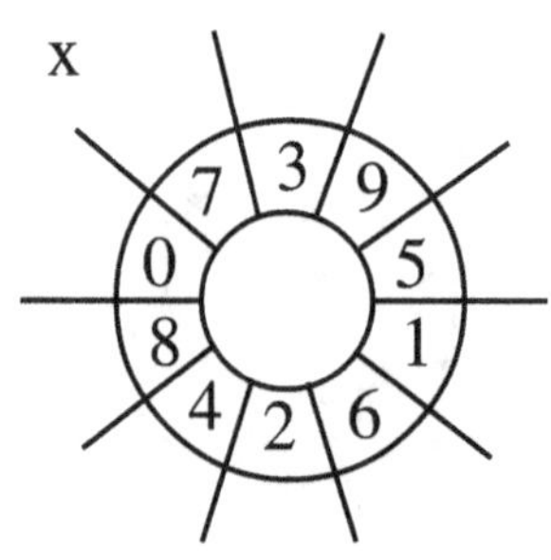

Review Exercises

1. 48
   $+ 73$

2. $76 - 25 =$

3. $372 + 26 =$

4. 758
   $- 423$

HELPFUL HINTS

1. Line up the numbers on the right.
2. Subtract the ones first.
3. It may be necessary to regroup more than once.

Examples:

$$\begin{array}{r} 2\,0\,3 \\ -\ 5\,6 \\ \hline (1\,4\,7) \end{array} \qquad \begin{array}{r} 7\,0\,0 \\ -\,2\,3\,4 \\ \hline (4\,6\,6) \end{array}$$

1	2	3	4	5	6	7	8	9	10	Score

S. 70
   $- 26$

S. 30
   $- 16$

1. 70
   $- 26$

2. 300
   $- 76$

3. 502
   $-\ 65$

4. 307
   $-\ 169$

5. 500
   $- 276$

6. 90
   $- 76$

7. Subtract 35 from 60.

8. $320 - 36 =$

9. 47
   $- 24$

10. 506
    $- 234$

Susan earned 50 dollars and Phil earned 42 dollars. How much more than Phil did Susan earn?

Problem Solving

12

It is illegal to photocopy this page. Copyright © 2023, Richard W. Fisher

Review Exercises	Speed Drills

Review Exercises

1. 60
    − 27

2. 975
    − 232

3. 90
    − 26

4. 71 − 63 =

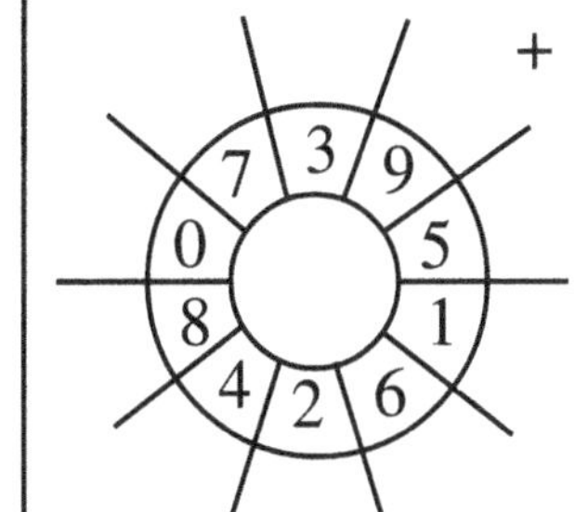

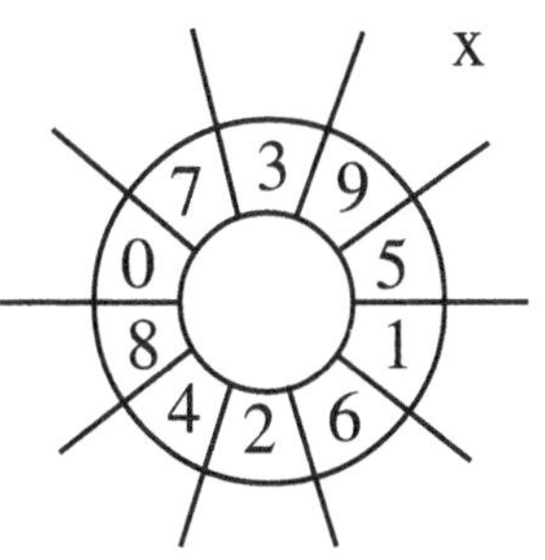

Use what you have learned to solve the following problems.

S. 732
    + 222

S. 342
    − 27

1. 42
    + 36

2. 543
    − 162

3. 7,231
    + 422

4. 761
    + 243

5. 80
    − 16

6. 765
    − 242

7. 70 − 23 =

8. Find the sum of 16 and 47.

9. Find the difference between 125 and 24.

10. 726 + 233 =

1	
2	
3	
4	
5	
6	
7	
8	
9	
10	
Score	

Problem Solving   A theater has 275 seats. If 243 have been taken, how many seats are left empty?

13

It is illegal to photocopy this page. Copyright © 2023, Richard W. Fisher

Speed Drills	Review Exercises

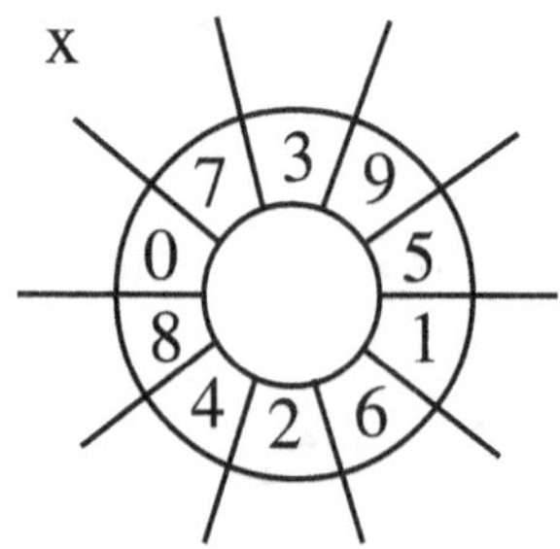

1. 302
    - 68

2. 17 + 23 + 24 =

3. Find the difference between 452 and 215.

4. 60 – 28 =

HELPFUL HINTS

1. Line the numbers up on the right.
2. Multiply the ones first.
3. Regroup when necessary.
4. "Product" means to multiply.

Examples:

$$\begin{array}{r} \overset{1}{2}\,4 \\ \times\,3 \\ \hline 72 \end{array} \qquad \begin{array}{r} \overset{1}{2}\overset{2}{3}\,6 \\ \times\,\;4 \\ \hline 944 \end{array}$$

1	
2	
3	
4	
5	
6	
7	
8	
9	
10	
Score	

S. 35
    x 3

S. 432
    x 6

1. 43
    x 3

2. 25
    x 6

3. 232
    x   4

4. 36
    x 4

5. 93
    x 3

6. 325
    x 7

7. 4 x 260 =

8. 5 x 272 =

9. Find the product of 3 and 252.

10. Multiply 7 and 125.

14

I have 7 boxes of crayons. Each box should have 14 crayons in it. How many crayons should I have altogether?

Problem Solving

It is illegal to photocopy this page. Copyright © 2023, Richard W. Fisher

Review Exercises	Speed Drills

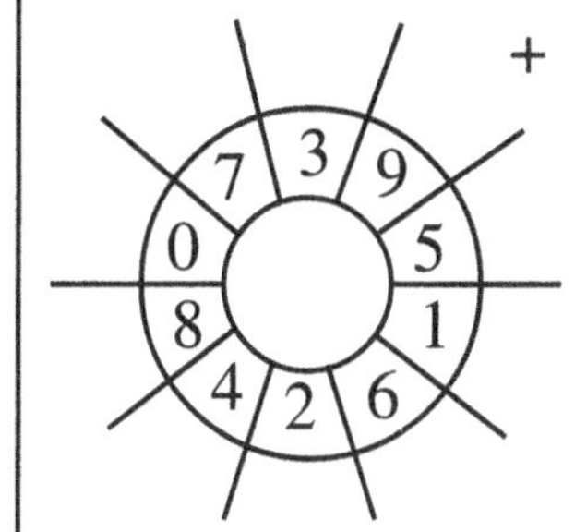

1. 23
    x 4

2. 423
    x 2

3. 97
    − 23

4. 37
    + 426

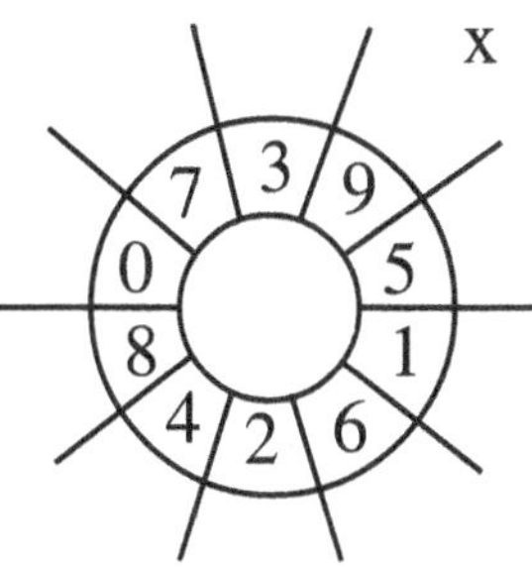

1. Line the numbers up on the right.   Examples:
2. Multiply the ones first.
3. Multiply the tens second.
4. Add the two products.
3. Remember to regroup whe
necessary.

```
   4 3        5 3 7
 x 3 2      x  2 4
 -----      -------
   8 6       2 1 4 8
 1 2 9 0    1 0 7 4 0
 -------    ---------
  1,3 7 6    1 2,8 8 8
```

S. 46
    x 23

S. 146
    x 42

1. 16
    x 12

2. 43
    x 12

3. 42
    x 13

4. 124
    x 23

5. 24
    x 30

6. 305
    x 23

7. Find the product of 20 and 24.

8. 16 x 232 =

9. Find the product of 31 and 132.

10. 33 x 120 =

1	
2	
3	
4	
5	
6	
7	
8	
9	
10	
Score	

Problem Solving | A small school has only 6 classrooms. There are 32 desks in each classroom. How many desks are in the school?

15

It is illegal to photocopy this page. Copyright © 2023, Richard W. Fisher

Speed Drills	Review Exercises

Speed Drills

+

x

Review Exercises

1. 304
    x   6

2. Find the product of 5 and 180.

3.   35
    + 216

4. Find the difference between 712 and 21.

HELPFUL HINTS

Use what you have learned to solve the following problems.

1	
2	
3	
4	
5	
6	
7	
8	
9	
10	
Score	

S.   23   S.   432   1.   26   2.   22
   x   6   x   3   x   3   x   6

3.   527   4.   47   5   47   6.   46
   x   3   x 30   x   4   x   23

7. Find the product of 4 and 16.

8. 30 x 12 =

9. 3 x 215 =

10. 432 x 23 =

16

Each package of paper contains 500 sheets of paper. How many sheets of paper do 3 packages contain?

Problem Solving

It is illegal to photocopy this page. Copyright © 2023, Richard W. Fisher

	Review Exercises	Speed Drills

1.   46
   − 25

2.   312
   x   2

3.   653
   + 374

4.   Find the product of 7 and 14.

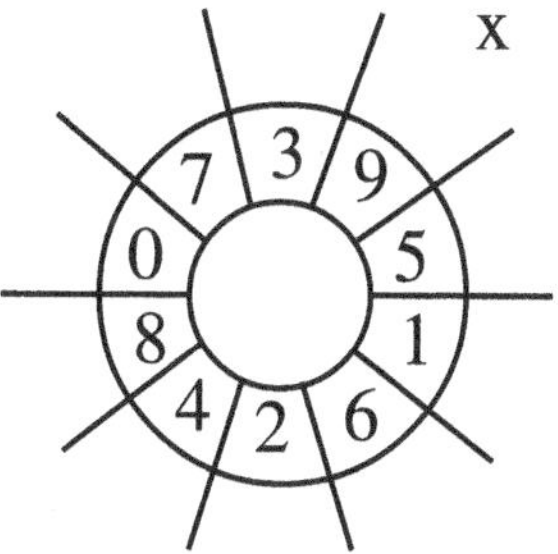

1. Divide
2. Multiply
3. Subtract
4. Begin again

Examples

    1 7 r 1
2) 3 5
   − 2 ↓
    1 5
   − 1 4
      1

    6 r 3
4) 2 7
  − 2 4
     3

Remember! The remainder must always be smaller than the divisor!

HELPFUL HINTS

S. 2) 37   S. 3) 69   1. 3) 43   2. 8) 43

3. 7) 87   4. 4) 93   5. 8) 97   6. 6) 43

7.   5) 66

8.   4) 97

9.   5) 61

10.   2) 84

1	
2	
3	
4	
5	
6	
7	
8	
9	
10	
Score	

Problem Solving | A teacher needs 72 rulers for his class. If rulers come in boxes that contain 6 rulers, how many boxes does the teacher need?

17

It is illegal to photocopy this page. Copyright © 2023, Richard W. Fisher

Speed Drills

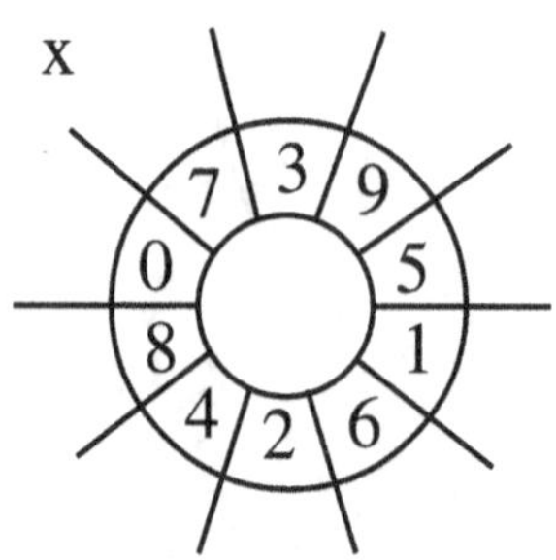

HELPFUL HINTS

Review Exercises

1. $2\overline{)35}$ 2. $6\overline{)67}$

3. $4 \times 236 =$ 4. $70 - 23 =$

1. Divide
2. Multiply
3. Subtract
4. Begin Again

Remember! The remainder must always be smaller than the divisor!

Examples:

$$
\begin{array}{r}
171 \ r2 \\
3\overline{)515} \\
-3\downarrow \\
\hline
21\downarrow \\
-21 \\
\hline
05 \\
-3 \\
\hline
2
\end{array}
$$

$$
\begin{array}{r}
203 \\
4\overline{)812} \\
-8\downarrow \\
\hline
01\downarrow \\
-0 \\
\hline
12 \\
-12 \\
\hline
0
\end{array}
$$

$$
\begin{array}{r}
67 \ r1 \\
5\overline{)336} \\
-30\downarrow \\
\hline
36 \\
-35 \\
\hline
1
\end{array}
$$

S. $3\overline{)435}$ S. $3\overline{)963}$ 1. $2\overline{)512}$

2. $2\overline{)624}$ 3. $7\overline{)77}$ 4. $5\overline{)412}$

5. $2\overline{)824}$ 6. $3\overline{)663}$ 7. $4\overline{)484}$

8. $4\overline{)88}$ 9. $2\overline{)48}$ 10. $6\overline{)606}$

Score
1
2
3
4
5
6
7
8
9
10

18

A theater has 45 seats. The seats are placed in 5 equal rows. How many seats are in each row?

Problem Solving

It is illegal to photocopy this page. Copyright © 2023, Richard W. Fisher

Speed Drills	Review Exercises

Speed Drills

+

7 3 9
0   5
8   1
4 2 6

x

7 3 9
0   5
8   1
4 2 6

HELPFUL HINTS ➤

1. Divide
2. Multiply
3. Subtract
4. Begin Again

Review Exercises

1. $5 \overline{)123}$ 2. $\begin{array}{r} 213 \\ \times \ \ 7 \\ \hline \end{array}$

3. $\begin{array}{r} 710 \\ - \ 167 \\ \hline \end{array}$ 4. $344 + 16 + 245 =$

Use what you have learned to solve the following problems.
Remember! Remainders must always be less than the divisor.
Zeroes may sometimes appear in the quotient.

S. $3 \overline{)24}$ S. $8 \overline{)568}$ 1. $2 \overline{)32}$

2. $5 \overline{)750}$ 3. $3 \overline{)765}$ 4. $5 \overline{)173}$

5. $6 \overline{)99}$ 6. $8 \overline{)49}$ 7. $3 \overline{)636}$

8. $3 \overline{)732}$ 9. $6 \overline{)606}$ 10. $7 \overline{)847}$

1	
2	
3	
4	
5	
6	
7	
8	
9	
10	
Score	

Problem Solving   Four boys worked together raking leaves. They earned a total of 200 dollars. If they shared the money equally, how much did each boy get?

It is illegal to photocopy this page. Copyright © 2023, Richard W. Fisher

1	
2	
3	
4	
5	
6	
7	
8	
9	
10	
11	
12	
13	
14	
15	
16	
17	
18	
19	
20	

1.　　23　　　　　2.　　324　　　　　3.　　$223 + 414 =$
　　$+\ 34$　　　　　　　$+\ \ 35$

5.　　342
　　　215
　$+\ \ 325$

4.　　$152 + 246 =$

6.　　35　　　　　7.　　732　　　　　8.　　$90 - 68 =$
　$-\ 14$　　　　　　$-\ 227$

9.　　$3{,}234 - 316 =$　　　　　10.　　$375 - 234 =$

11.　34　　　　　12.　423　　　　　13.　　123
　$x\ \ 2$　　　　　　　$x\ \ 4$　　　　　　　$x\ \ 7$

14.　26　　　　　15.　213
　$x\ 12$　　　　　　$x\ 23$　　　　　16.　$2\,\overline{)\,27}$

17.　$4\,\overline{)\,55}$　　　　　　　18.　$5\,\overline{)\,167}$

19.　$3\,\overline{)\,642}$　　　　　　　20.　$2\,\overline{)\,672}$

20

It is illegal to photocopy this page. Copyright © 2023, Richard W. Fisher

Speed Drills	Review Exercises

+

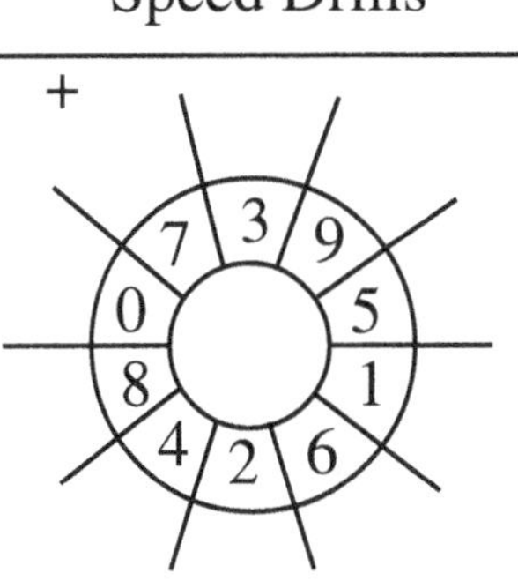

1.   70
    - 16

2.   23
    + 24

3. $3\overline{)342}$

4.   224
    x 3

x

Example:

A fraction is a number that names a part of a whole or a group.

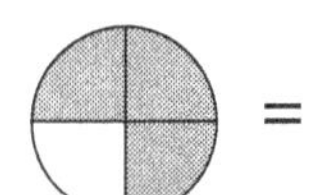 $= \dfrac{3}{4} \rightarrow$ numerator, denominator

Think of $\dfrac{3}{4}$ as $\dfrac{3 \text{ of}}{4 \text{ equal parts}}$

HELPFUL HINTS 

Write a fraction for each shaded figure (some may have more than one name).

| | 1 |
| 2 |
| 3 |
| 4 |
| 5 |
| 6 |
| 7 |
| 8 |
| 9 |
| 10 |
| Score |

S. 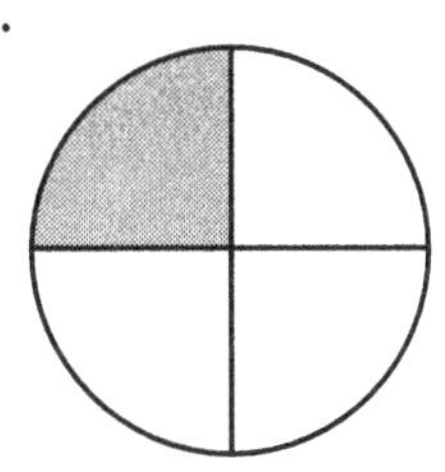   S. 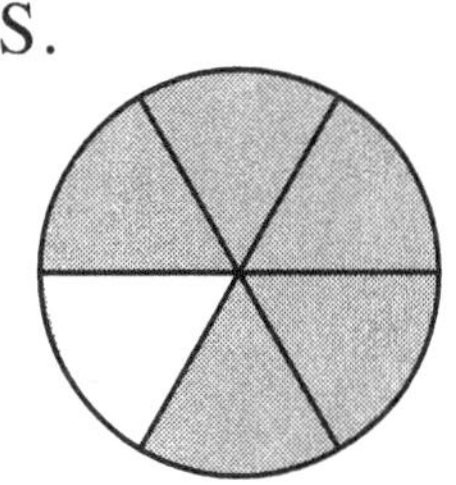   1. 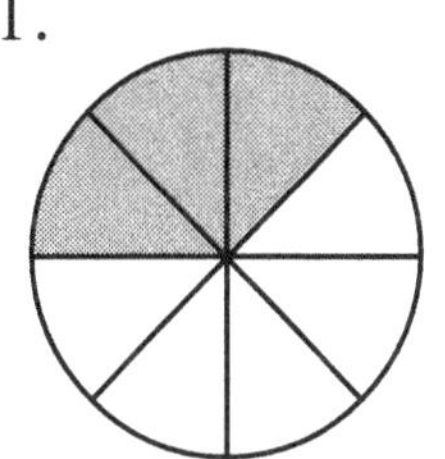   2. 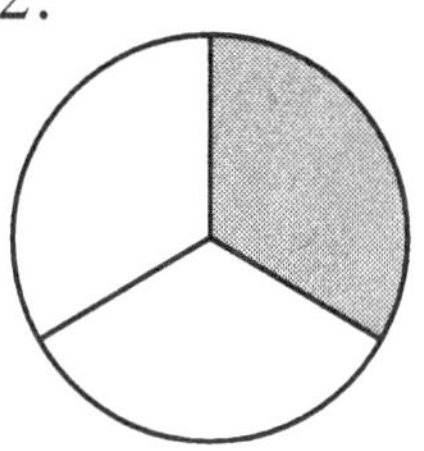

3. 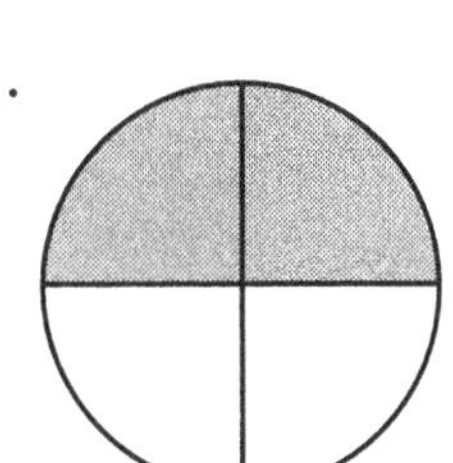   4. 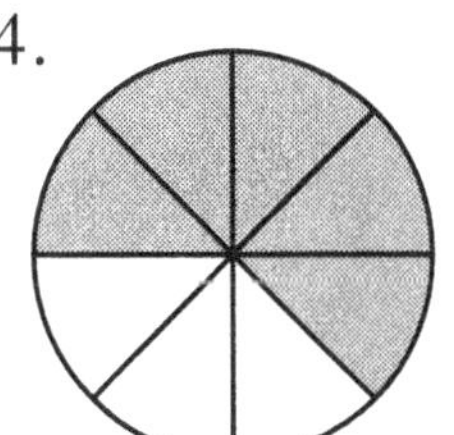   5. 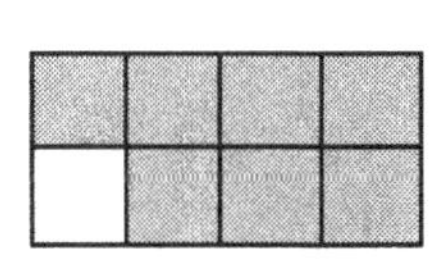   6. 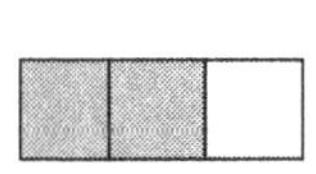

7. 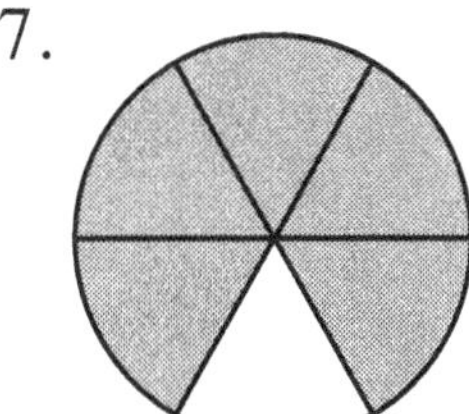   8. 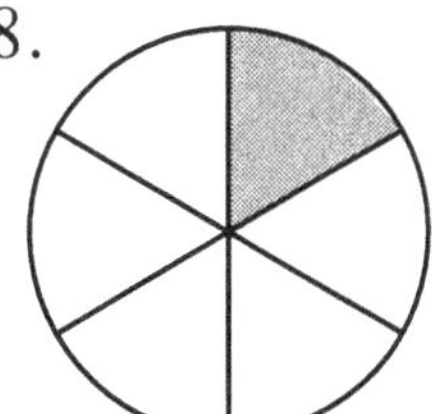   9. 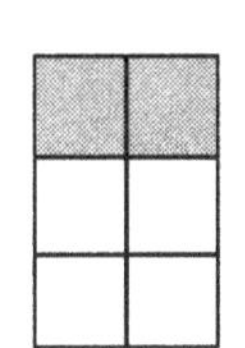   10. 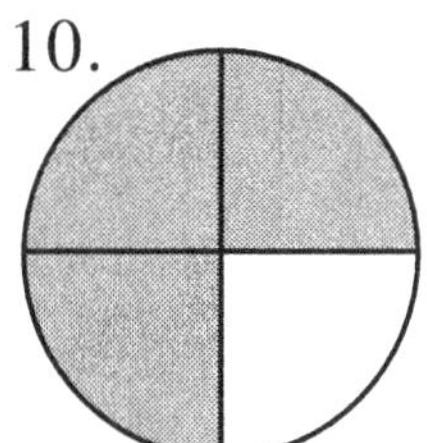

Extra credit: On a separate sheet of paper draw a figure for the following fractions. $\dfrac{1}{2}, \dfrac{1}{4}, \dfrac{1}{8}, \dfrac{3}{8}, \dfrac{2}{3}, \dfrac{5}{6}$

Problem Solving

6 boxes weigh a total of 30 pounds. If each box weighs the same, how much does each box weigh?

21

It is illegal to photocopy this page. Copyright © 2023, Richard W. Fisher

| Review Exercises | Speed Drills |

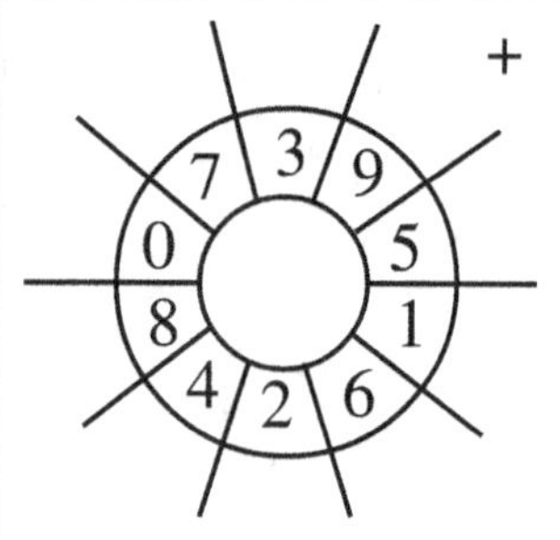

1. 4 x 213 =   2. 16 + 223 =

3.   525   4. 2) 428
    − 14

$\frac{2}{4}$ has been reduced to its simplest form, which is $\frac{1}{2}$

Divide the numerator and the denominator by the largest possible number.

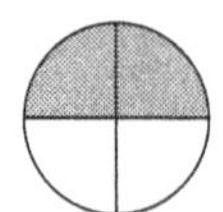 = $\frac{2}{4}$ = $\frac{1}{2}$

Examples

$2 \overline{) \frac{6}{8}} = \left(\frac{3}{4}\right)$

$5 \overline{) \frac{5}{10}} = \left(\frac{1}{2}\right)$

$2 \overline{) \frac{4}{6}} = \left(\frac{2}{3}\right)$

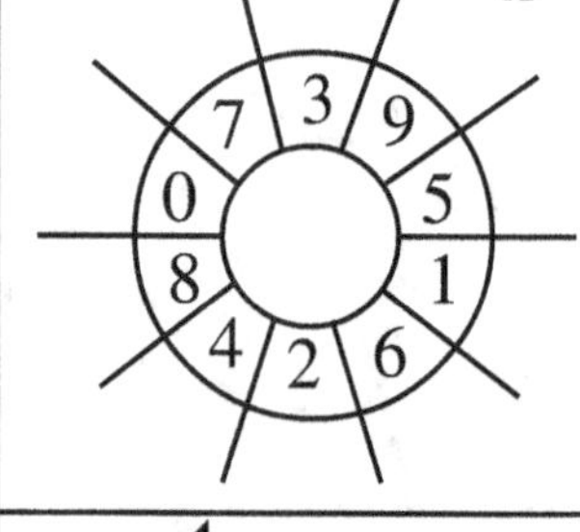

HELPFUL HINTS

Reduce each fraction to its lowest terms.

S. $\frac{3}{6}$ =   S. $\frac{2}{8}$ =   1. $\frac{2}{10}$ =

2. $\frac{2}{6}$ =   3. $\frac{6}{9}$ =   4. $\frac{10}{15}$ =

5. $\frac{8}{10}$ =   6. $\frac{3}{9}$ =   7. $\frac{5}{15}$ =

8. $\frac{2}{12}$ =   9. $\frac{6}{10}$ =   10. $\frac{7}{14}$ =

1	
2	
3	
4	
5	
6	
7	
8	
9	
10	
Score	

22

If there are 12 crayons in each box, how many crayons are there in 3 boxes?

Problem Solving

It is illegal to photocopy this page. Copyright © 2023, Richard W. Fisher

Speed Drills

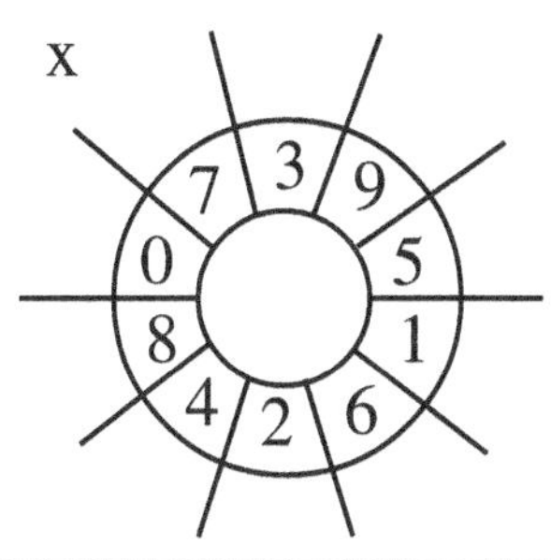

+

x

HELPFUL
HINTS

Review Exercises

1. What fraction of the figure is shaded? 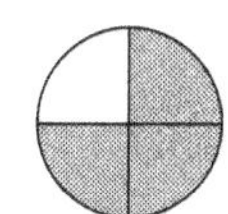

2. Reduce $\frac{4}{6}$ to its lowest terms.

3. $2 \overline{)\ 428}$

4. $\begin{array}{r} 163 \\ \times\ \ 5 \\ \hline \end{array}$

An improper fraction has a numerator that is equal to or greater than its denominator. Improper fractions can be written either as whole numbers or as mixed numerals (a whole number and a fraction). To change, divide the numerator by the denominator. Example:

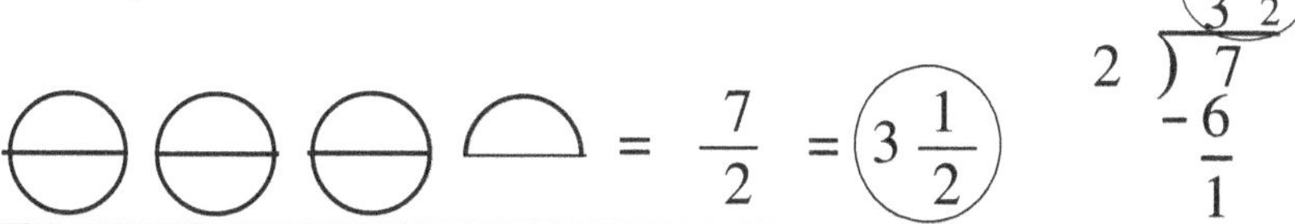

$$= \frac{7}{2} = 3\frac{1}{2}$$

$$2 \overline{)\ 7} \quad 3\frac{1}{2}$$
$$-6$$
$$\overline{\ \ 1}$$

Change each improper fraction to a mixed number or a whole number.

S. $\frac{3}{2} =$ S. $\frac{7}{6} =$ 1. $\frac{7}{4} =$

2. $\frac{5}{2} =$ 3. $\frac{8}{5} =$ 4. $\frac{10}{7} =$

5. $\frac{6}{5} =$ 6. $\frac{4}{3} =$ 7. $\frac{12}{5} =$

8. $\frac{7}{3} =$ 9. $\frac{11}{5} =$ 10. $\frac{8}{3} =$

1	
2	
3	
4	
5	
6	
7	
8	
9	
10	
Score	

Problem Solving

On Saturday 120 people went dancing and on Sunday 96 people went dancing. How many more Saturday dancers were there than Sunday dancers?

23

It is illegal to photocopy this page. Copyright © 2023, Richard W. Fisher

Review Exercises	Speed Drills

1. Change $\frac{6}{5}$ to a mixed numeral.

2. Change $\frac{10}{7}$ to a mixed numeral.

3. Reduce $\frac{4}{6}$ to its lowest terms.

4. What fraction of the figure is shaded?

To add fractions with like denominators, first add the numerators, then ask the following questions about your answer:
1. Is the answer an improper fraction? If it is, convert it to a mixed numberal or whole number.
2. Can the fraction be reduced? If it can be, reduce it to its simplest form.

Examples:

$$\begin{array}{r}\frac{1}{5}\\+\frac{2}{5}\\\hline \left(\frac{3}{5}\right)\end{array} \qquad \begin{array}{r}\frac{1}{8}\\+\frac{3}{8}\\\hline \frac{4}{8}=\left(\frac{1}{2}\right)\end{array} \qquad \begin{array}{r}\frac{3}{4}\\+\frac{3}{4}\\\hline \frac{6}{4}=1\frac{2}{4}\\=\left(1\frac{1}{2}\right)\end{array}$$

HELPFUL HINTS

S. $\begin{array}{r}\frac{1}{5}\\\frac{3}{5}\\+\\\hline\end{array}$ S. $\begin{array}{r}\frac{5}{6}\\\frac{3}{6}\\+\\\hline\end{array}$ 1. $\begin{array}{r}\frac{2}{7}\\\frac{3}{7}\\+\\\hline\end{array}$ 2. $\begin{array}{r}\frac{4}{7}\\\frac{5}{7}\\+\\\hline\end{array}$

3. $\begin{array}{r}\frac{3}{5}\\\frac{3}{5}\\+\\\hline\end{array}$ 4. $\begin{array}{r}\frac{1}{8}\\\frac{5}{8}\\+\\\hline\end{array}$ 5. $\begin{array}{r}\frac{5}{6}\\\frac{1}{6}\\+\\\hline\end{array}$ 6. $\begin{array}{r}\frac{7}{8}\\\frac{2}{8}\\+\\\hline\end{array}$

7. $\begin{array}{r}\frac{3}{8}\\\frac{1}{8}\\+\\\hline\end{array}$ 8. $\begin{array}{r}\frac{7}{10}\\\frac{1}{10}\\+\\\hline\end{array}$ 9. $\begin{array}{r}\frac{7}{10}\\\frac{3}{10}\\+\\\hline\end{array}$ 10. $\begin{array}{r}\frac{1}{3}\\\frac{1}{3}\\+\\\hline\end{array}$

	Score
1	
2	
3	
4	
5	
6	
7	
8	
9	
10	

24

Problem Solving

If $\frac{1}{8}$ of the kids ride their bikes to school and $\frac{3}{8}$ of them walk, what fraction of them walk or ride their bikes to school?

It is illegal to photocopy this page. Copyright © 2023, Richard W. Fisher

Speed Drills

+

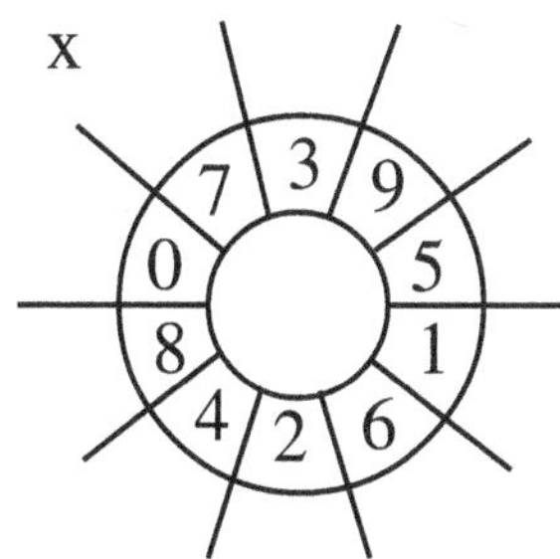

x

HELPFUL HINTS

Review Exercises

1. $\dfrac{2}{5}$ $+ \dfrac{1}{5}$

2. $3\overline{)69}$

3. $\dfrac{7}{10}$ $+ \dfrac{5}{10}$

4. $7 + 14 + 12 =$

Examples:

1. Add the fractions first.
2. Add the whole numbers next.
3. If there is an improper fraction, change it to a mixed numeral.
4. Add the mixed numeral to the whole number.

*Reduce fractions to lowest terms

$3\frac{1}{8}$
$+ 2\frac{3}{8}$
$5\frac{4}{8} = \boxed{5\frac{1}{2}}$

$2\frac{3}{5}$
$+ 3\frac{4}{5}$
$5\frac{7}{5} = 5 + 1\frac{2}{5} = \boxed{6\frac{2}{5}}$

$3\frac{5}{8}$
$+ 2\frac{5}{8}$
$5\frac{10}{8} = 5 + 1\frac{2}{8} = 6\frac{2}{8} = \boxed{6\frac{1}{4}}$

1	
2	
3	
4	
5	
6	
7	
8	
9	
10	
Score	

S. $3\frac{1}{4}$ $+ 2\frac{1}{4}$

S. $3\frac{4}{5}$ $+ 2\frac{2}{5}$

1. $3\frac{1}{5}$ $+ 2\frac{2}{5}$

2. $2\frac{1}{8}$ $+ 3\frac{1}{8}$

3. $2\frac{1}{6}$ $+ 3\frac{2}{6}$

4. $3\frac{1}{10}$ $+ 2\frac{3}{10}$

5. $3\frac{4}{7}$ $+ 2\frac{4}{7}$

6. $3\frac{1}{6}$ $+ 2\frac{2}{6}$

7. $4\frac{3}{8}$ $+ 2\frac{1}{8}$

8. $2\frac{1}{6}$ $+ 3\frac{3}{6}$

9. $3\frac{1}{5}$ $+ 2\frac{2}{5}$

10. $2\frac{3}{4}$ $+ 2\frac{1}{4}$

Problem Solving

Juan bakes a pie and a cake. He uses $\frac{2}{9}$ cups of flour for the cake and $\frac{1}{9}$ cups for the pie crust. How much flour did he use?

25

It is illegal to photocopy this page. Copyright © 2023, Richard W. Fisher

| | Review Exercises | Speed Drills |

Review Exercises

1. Reduce $\frac{6}{8}$ to its lowest terms.

2. Change $\frac{5}{2}$ to a mixed numeral.

3. $\begin{array}{r} \frac{2}{5} \\ + \frac{2}{5} \\ \hline \end{array}$ 4. $\begin{array}{r} 2\frac{3}{5} \\ + 1\frac{3}{5} \\ \hline \end{array}$

Speed Drills

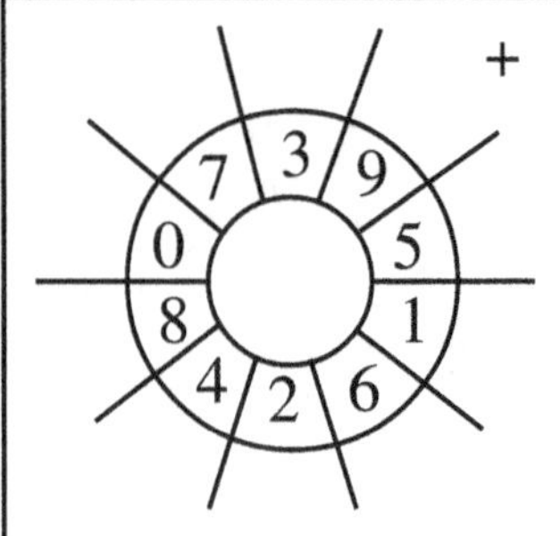

To subtract fractions that have like denominators, first subtract the numerators, then, **if necessary, reduce the answer to its lowest terms.**

Examples:

$\begin{array}{r} \frac{4}{5} \\ - \frac{1}{5} \\ \hline \frac{3}{5} \end{array}$ $\begin{array}{r} \frac{5}{6} \\ - \frac{1}{6} \\ \hline \frac{4}{6} \end{array} = \frac{2}{3}$

HELPFUL HINTS

S. $\begin{array}{r} \frac{3}{8} \\ - \frac{1}{8} \\ \hline \end{array}$ S. $\begin{array}{r} \frac{3}{4} \\ - \frac{1}{4} \\ \hline \end{array}$ 1. $\begin{array}{r} \frac{5}{8} \\ - \frac{1}{8} \\ \hline \end{array}$ 2. $\begin{array}{r} \frac{3}{6} \\ - \frac{1}{6} \\ \hline \end{array}$

3. $\begin{array}{r} \frac{5}{7} \\ - \frac{2}{7} \\ \hline \end{array}$ 4. $\begin{array}{r} \frac{9}{10} \\ - \frac{1}{10} \\ \hline \end{array}$ 5. $\begin{array}{r} \frac{7}{11} \\ - \frac{4}{11} \\ \hline \end{array}$ 6. $\begin{array}{r} \frac{6}{7} \\ - \frac{1}{7} \\ \hline \end{array}$

7. $\begin{array}{r} \frac{7}{10} \\ - \frac{3}{10} \\ \hline \end{array}$ 8. $\begin{array}{r} \frac{7}{8} \\ - \frac{3}{8} \\ \hline \end{array}$ 9. $\begin{array}{r} \frac{2}{3} \\ - \frac{1}{3} \\ \hline \end{array}$ 10. $\begin{array}{r} \frac{7}{9} \\ - \frac{1}{9} \\ \hline \end{array}$

1	
2	
3	
4	
5	
6	
7	
8	
9	
10	
Score	

26 Bella lives $\frac{4}{5}$ of a mile from school. If she has already walked $\frac{3}{5}$ of a mile, how much farther does she have to go?

Problem Solving

It is illegal to photocopy this page. Copyright © 2023, Richard W. Fisher

Speed Drills	Review Exercises

Review Exercises

1. $\frac{1}{4}$ 2. $\frac{4}{5}$ 3. $3\frac{3}{7}$ 4. $3\frac{4}{5}$

$+\ \frac{1}{4}$ $+\ \frac{3}{5}$ $+\ 2\frac{2}{7}$ $+\ 2\frac{2}{5}$

HELPFUL HINTS

To subtract a fraction or a mixed number from a whole number, take one from the whole number and make it a fraction, then subtract.

Example:

$$7 \;=\; \overset{6}{\cancel{7}} \to \frac{5}{5}$$

$$-\ 2\frac{3}{5} \qquad -\ 2\ \frac{3}{5}$$

$$\boxed{4\frac{2}{5}}$$

S. 6   S. 7   1. 6   2. 5

$-\ 2\frac{3}{5}$ $-\ \frac{3}{4}$ $-\ 2\frac{4}{7}$ $-\ 1\frac{3}{5}$

3. 7   4. 6   5. 7   6. 9

$-\ \frac{2}{3}$ $-\ 2\frac{9}{10}$ $-\ 2\frac{1}{8}$ $-\ 2\frac{3}{7}$

7. 7   8. 4   9. 6   10. 5

$-\ 3\frac{7}{9}$ $-\ 3\frac{1}{2}$ $-\ 2\frac{3}{10}$ $-\ \frac{3}{5}$

	Score
1	
2	
3	
4	
5	
6	
7	
8	
9	
10	

Problem Solving

A tailor has 5 yards of cloth. If he uses 2 yards to make a shirt, how many yards does he have left?

27

It is illegal to photocopy this page. Copyright © 2023, Richard W. Fisher

Review Exercises	Speed Drills

1. $\frac{2}{3}$ $-\frac{1}{3}$ 2. $\frac{3}{4}$ $-\frac{1}{4}$ 3. $2\frac{1}{4}$ $+3\frac{1}{4}$ 4. 199 -128

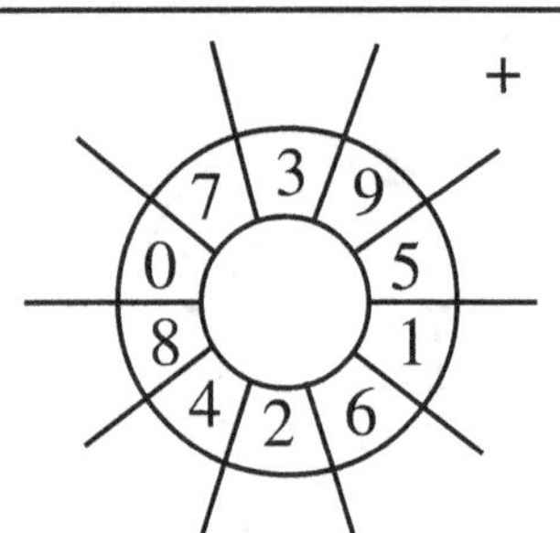

To subtract mixed numbers with like denominators, subtract the fractions first, then the whole numbers. Reduce the fractions in the answer to lowest terms. If the fractions can't be subtracted as they are written, take one from the whole number and increase the fraction, then subtract.

$3\frac{3}{4}$
$-1\frac{1}{4}$
$2\frac{2}{4} = \left(2\frac{1}{2}\right)$

$\cancel{3}4\frac{1}{4} + \frac{4}{4} = \frac{5}{4}$
$-2\frac{3}{4}$
$1\frac{2}{4} = \left(1\frac{1}{2}\right)$

HELPFUL HINTS

S. $3\frac{3}{4}$ $-1\frac{1}{4}$

S. $5\frac{2}{3}$ $-2\frac{2}{3}$

1. $3\frac{3}{5}$ $-1\frac{2}{5}$

2. $4\frac{3}{6}$ $-1\frac{1}{6}$

3. $5\frac{5}{6}$ $-2\frac{1}{6}$

4. $7\frac{7}{10}$ $-2\frac{2}{10}$

5. $3\frac{4}{5}$ $-1\frac{2}{5}$

6. $4\frac{3}{4}$ $-2\frac{1}{4}$

7. $5\frac{1}{7}$ $-2\frac{1}{7}$

8. $3\frac{7}{8}$ $-1\frac{5}{8}$

9. $7\frac{1}{3}$ $-3\frac{1}{3}$

10. $3\frac{4}{5}$ $-2\frac{1}{5}$

1	
2	
3	
4	
5	
6	
7	
8	
9	
10	
Score	

28 A woman worked $1\frac{2}{3}$ hours on Monday and $3\frac{1}{3}$ hours on Tuesday. How many hours did she work in all?

Problem Solving

It is illegal to photocopy this page. Copyright © 2023, Richard W. Fisher

Speed Drills	Review Exercises

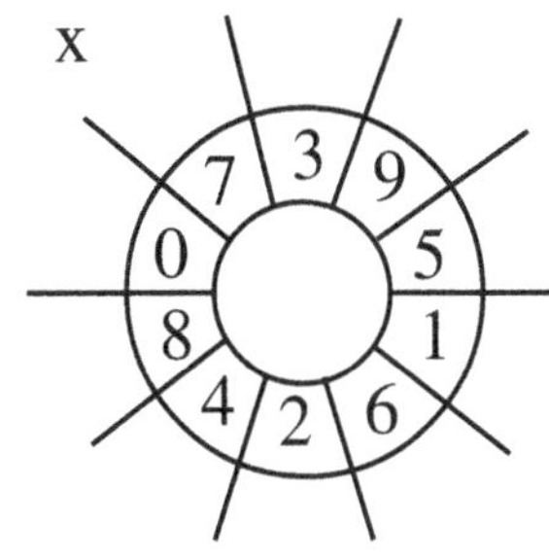

Review Exercises

1. $\dfrac{3}{10}$
$-\dfrac{1}{10}$

2. $\dfrac{3}{4}$
$+\dfrac{1}{4}$

3. Change $\dfrac{8}{3}$ to a mixed numeral.

4. Reduce $\dfrac{3}{9}$ to its lowest terms.

Use what you have learned to solve the following problems. Regroup when necessary.

Reduce all answers to lowest terms.

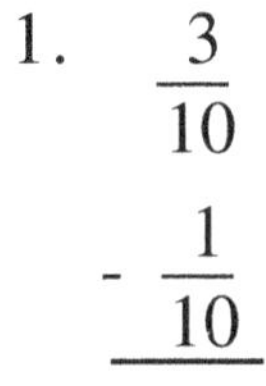

S. $2\dfrac{1}{6}$
$+\ 3\dfrac{2}{6}$

S. $5\dfrac{1}{4}$
$-\ 2\dfrac{3}{4}$

1. $\dfrac{2}{5}$
$+\ \dfrac{1}{5}$

2. $\dfrac{5}{8}$
$+\ \dfrac{1}{8}$

3. $\dfrac{8}{9}$
$+\ \dfrac{2}{9}$

4. $\dfrac{7}{8}$
$-\ \dfrac{5}{8}$

5. 6
$-\ 2\dfrac{1}{4}$

6. 5
$-\ \dfrac{2}{3}$

7. $3\dfrac{9}{10}$
$+\ 2\dfrac{3}{10}$

8. $6\dfrac{1}{3}$
$-\ 2\dfrac{2}{3}$

9. $3\dfrac{1}{5}$
$-\ 2\dfrac{3}{5}$

10. $4\dfrac{1}{6}$
$-\ 2\dfrac{4}{6}$

	1
	2
	3
	4
	5
	6
	7
	8
	9
	10
	Score

Problem Solving

A family has $7\dfrac{1}{3}$ pounds of beef in the freezer. If they use $5\dfrac{2}{3}$ pounds for supper, how many pounds do they have left?

29

It is illegal to photocopy this page. Copyright © 2023, Richard W. Fisher

Review Exercises	Speed Drills

1. Find the sum of $\frac{3}{7}$ and $\frac{2}{7}$

2. Find the difference between $\frac{3}{4}$ and $\frac{1}{4}$

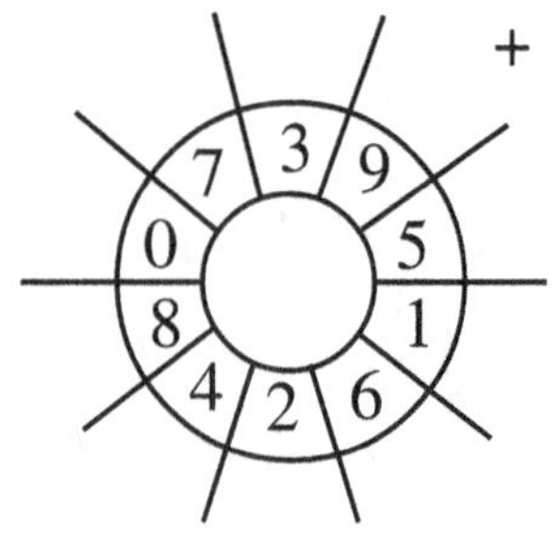

3. $\begin{array}{r} 7 \\ -\ 2\frac{1}{2} \\ \hline \end{array}$ 4. $\begin{array}{r} 5\frac{2}{3} \\ -\ 3 \\ \hline \end{array}$

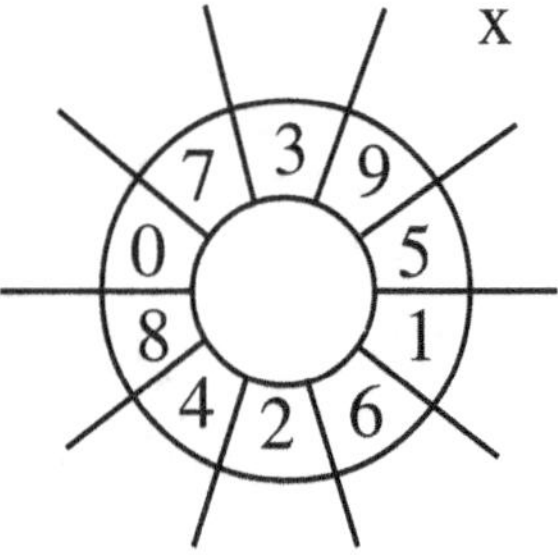

To add or subtract fractions with unlike denominators, you need to first find their **least common denominator** (LCD). The LCD is the smallest number, other than zero, that each denominator will divide into evenly. Examples: The LCD of

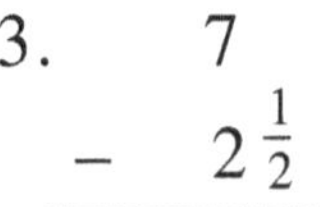

$\frac{1}{3}$ and $\frac{1}{2}$ is 6 $\frac{1}{5}$ and $\frac{1}{10}$ is 10 $\frac{1}{4}$ and $\frac{1}{6}$ is 12

HELPFUL HINTS

Find the least common denominators of each of the following:

S. $\frac{1}{3}$ and $\frac{1}{4}$ S. $\frac{5}{6}$ and $\frac{1}{8}$ 1. $\frac{1}{5}$ and $\frac{1}{2}$

2. $\frac{3}{4}$ and $\frac{1}{8}$ 3. $\frac{2}{3}$ and $\frac{1}{9}$ 4. $\frac{1}{5}$ and $\frac{1}{15}$

5. $\frac{4}{5}$ and $\frac{3}{2}$ 6. $\frac{3}{4}$ and $\frac{1}{16}$ 7. $\frac{4}{5}$ and $\frac{1}{4}$

8. $\frac{9}{10}$ and $\frac{1}{2}$ 9. $\frac{3}{14}$ and $\frac{1}{7}$ 10. $\frac{2}{5}$ and $\frac{1}{6}$

1	
2	
3	
4	
5	
6	
7	
8	
9	
10	
Score	

Problem Solving

30   A plane can travel 600 miles in one hour. If the speed remains the same, how far can it travel in 2 hours?

It is illegal to photocopy this page. Copyright © 2023, Richard W. Fisher

Speed Drills

+

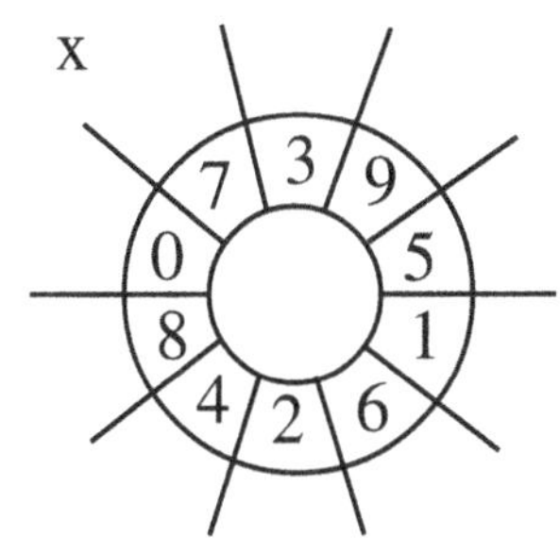

x

HELPFUL HINTS

Review Exercises

1. $3\overline{)603}$

2. $\begin{array}{r} 23 \\ \times\ 24 \\ \hline \end{array}$

3. $36 + 4 + 213 =$

4. $60 - 39 =$

To add or subtract fractions with unlike denominators, find the least common denominator. Multiply each fraction by one to make equivalent fractions. Finally, add or subtract.

Examples:

$$\frac{2}{3} \times \frac{2}{2} = \frac{4}{6}$$
$$-\ \frac{1}{2} \times \frac{3}{3} = \frac{3}{6}$$
$$\overline{\qquad\left(\frac{1}{6}\right)}$$

$$\frac{1}{2} \times \frac{4}{4} = \frac{4}{8}$$
$$+\ \frac{7}{8} \qquad = \frac{7}{8}$$
$$\overline{\qquad \frac{11}{8} = \left(1\frac{3}{8}\right)}$$

***Be sure to reduce all fractions to lowest terms.**

	Score
1	
2	
3	
4	
5	
6	
7	
8	
9	
10	

S. $\quad \dfrac{1}{4}$
$+ \ \dfrac{1}{3}$

S. $\quad \dfrac{4}{5}$
$- \ \dfrac{1}{2}$

1. $\quad \dfrac{2}{3}$
$+ \ \dfrac{1}{4}$

2. $\quad \dfrac{2}{3}$
$- \ \dfrac{1}{2}$

3. $\quad \dfrac{4}{5}$
$+ \ \dfrac{1}{2}$

4. $\quad \dfrac{2}{3}$
$+ \ \dfrac{1}{9}$

5. $\quad \dfrac{1}{2}$
$- \ \dfrac{1}{3}$

6. $\quad \dfrac{3}{5}$
$- \ \dfrac{1}{10}$

7. $\quad \dfrac{2}{5}$
$+ \ \dfrac{1}{3}$

8. $\quad \dfrac{1}{4}$
$+ \ \dfrac{1}{2}$

9. $\quad \dfrac{7}{10}$
$- \ \dfrac{1}{5}$

10. $\quad \dfrac{2}{3}$
$+ \ \dfrac{1}{2}$

Problem Solving

John bought 7 gallons of paint to paint his house. If he used $5\frac{3}{8}$ gallons, how much paint does he have left?

31

It is illegal to photocopy this page. Copyright © 2023, Richard W. Fisher

Review Exercises	Speed Drills

1. $3\overline{)69}$

2. $\dfrac{7}{8}$
   $-\dfrac{1}{8}$

3. $\dfrac{1}{4}$
   $+\dfrac{1}{3}$

4. $\dfrac{2}{3}$
   $-\dfrac{1}{2}$

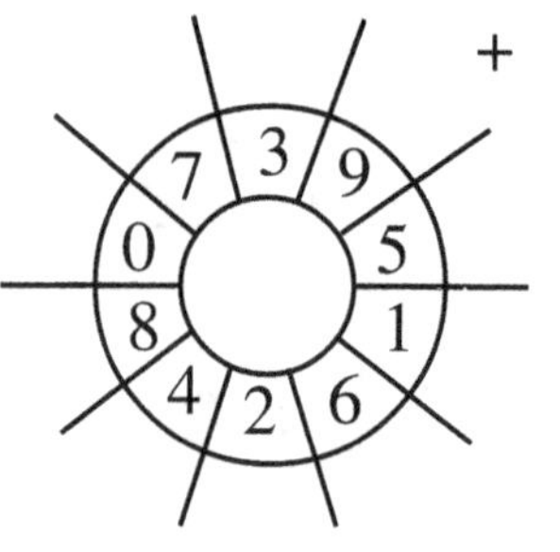

When adding mixed numerals with unlike denominators, first add the fractions. If the answer is an improper fraction, first make it into a mixed numeral, then add the sum to the the sum of the whole numbers. **Always reduce your answer to the lowest terms.**

Example:

$$3\tfrac{2}{3} \times \tfrac{2}{2} = \tfrac{4}{6}$$
$$+2\tfrac{1}{2} \times \tfrac{3}{3} = \tfrac{3}{6}$$
$$5 \qquad \tfrac{7}{6} = 1\tfrac{1}{6} = \left(6\tfrac{1}{6}\right)$$

S. $2\tfrac{2}{5}$
   $+\ 3\tfrac{1}{2}$

S. $4\tfrac{1}{2}$
   $+\ 3\tfrac{3}{5}$

1. $2\tfrac{1}{2}$
   $+\ 3\tfrac{1}{3}$

2. $2\tfrac{1}{5}$
   $+\ 3\tfrac{1}{2}$

3. $3\tfrac{1}{4}$
   $+\ 2\tfrac{2}{5}$

4. $2\tfrac{3}{4}$
   $+\ 1\tfrac{1}{3}$

5. $2\tfrac{1}{2}$
   $+\ 3\tfrac{1}{4}$

6. $5\tfrac{2}{3}$
   $+\ 2\tfrac{1}{6}$

7. $1\tfrac{1}{2}$
   $+\ 2\tfrac{1}{5}$

8. $3\tfrac{2}{5}$
   $+\ 2\tfrac{1}{3}$

9. $3\tfrac{1}{3}$
   $+\ 2\tfrac{2}{5}$

10. $3\tfrac{1}{5}$
    $+\ 2\tfrac{1}{4}$

	Score
1	
2	
3	
4	
5	
6	
7	
8	
9	
10	

A factory can produce 72 parts per hour. How many parts can be produced in 3 hours?

Problem Solving

It is illegal to photocopy this page. Copyright © 2023, Richard W. Fisher

Review Exercises	Speed Drills

1. Reduce $\frac{10}{15}$ to its lowest terms.

2. Change $\frac{12}{5}$ to a mixed numeral.

3. Find the least common denominator for $\frac{1}{2}$ and $\frac{2}{5}$.

4. $\begin{array}{r} \frac{1}{3} \\ + \frac{2}{5} \\ \hline \end{array}$

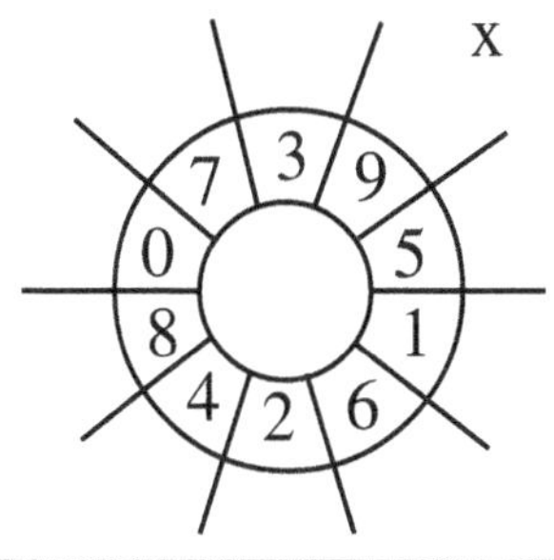

Use what you have learned to solve the following problems. **Remember to reduce all fractions to their lowest terms.**

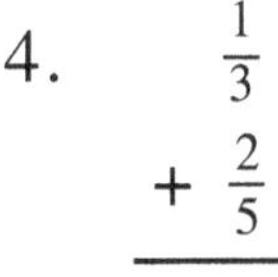

S. $\begin{array}{r} 3\frac{1}{2} \\ - 1\frac{1}{4} \\ \hline \end{array}$

S. $\begin{array}{r} 2\frac{1}{2} \\ + 2\frac{3}{4} \\ \hline \end{array}$

1. $\begin{array}{r} \frac{1}{5} \\ + \frac{3}{5} \\ \hline \end{array}$

2. $\begin{array}{r} \frac{2}{3} \\ + \frac{1}{4} \\ \hline \end{array}$

3. $\begin{array}{r} \frac{3}{5} \\ - \frac{1}{2} \\ \hline \end{array}$

4. $\begin{array}{r} 5 \\ - 2\frac{1}{5} \\ \hline \end{array}$

5. $\begin{array}{r} 3\frac{2}{3} \\ + 2\frac{2}{3} \\ \hline \end{array}$

6. $\begin{array}{r} 4\frac{1}{2} \\ - 1\frac{1}{5} \\ \hline \end{array}$

7. $\begin{array}{r} 3\frac{2}{3} \\ + 2\frac{1}{2} \\ \hline \end{array}$

8. $\begin{array}{r} 3\frac{1}{2} \\ - 2 \\ \hline \end{array}$

9. $\begin{array}{r} \frac{3}{8} \\ + \frac{1}{4} \\ \hline \end{array}$

10. $\begin{array}{r} 2\frac{1}{10} \\ + 2\frac{2}{5} \\ \hline \end{array}$

1	
2	
3	
4	
5	
6	
7	
8	
9	
10	
Score	

Problem Solving   Susan earned 7 dollars on Monday and 12 dollars on Tuesday. How much more did she earn on Tuesday than on Monday?

33

It is illegal to photocopy this page. Copyright © 2023, Richard W. Fisher

Speed Drills	Review Exercises

Speed Drills

+

x

HELPFUL HINTS

Review Exercises

1. $6 \overline{)72}$

2. $\begin{array}{r} 26 \\ \times\ 4 \\ \hline \end{array}$

3. $26 + 13 + 24 =$

4. $752 - 166 =$

When multiplying common fractions:
- First multiply the numerators.
- Next, multiply the denominators.
- If the answer is an improper fraction, change it to a mixed numeral.
- **Be sure to reduce fractions to their lowest terms.**
- **Remember: "of" means the same as "x" or "times."**

Examples

$$\frac{2}{3} \ \text{x} \ \frac{1}{2} = \frac{2}{6} = \boxed{\frac{1}{3}}$$

$$\frac{3}{2} \ \text{x} \ \frac{3}{4} = \frac{9}{8} = \boxed{1\frac{1}{8}}$$

1	
2	
3	
4	
5	
6	
7	
8	
9	
10	
Score	

S. $\dfrac{2}{5} \ \text{x} \ \dfrac{3}{5} =$

S. $\dfrac{4}{3} \ \text{x} \ \dfrac{4}{5} =$

1. $\dfrac{1}{2} \ \text{x} \ \dfrac{1}{3} =$

2. $\dfrac{2}{5} \ \text{x} \ \dfrac{1}{2} =$

3. $\dfrac{5}{2} \ \text{x} \ \dfrac{3}{5} =$

4. $\dfrac{3}{7} \ \text{x} \ \dfrac{5}{2} =$

5. $\dfrac{1}{2} \ \text{of} \ \dfrac{4}{5} =$

6. $\dfrac{2}{7} \ \text{x} \ \dfrac{3}{5} =$

7. $\dfrac{3}{2} \ \text{x} \ \dfrac{4}{5} =$

8. $\dfrac{2}{3} \ \text{of} \ \dfrac{4}{5} =$

9. $\dfrac{4}{3} \ \text{x} \ \dfrac{5}{6} =$

10. $\dfrac{1}{4} \ \text{of} \ \dfrac{3}{5} =$

34

Mom cooked a $3\frac{1}{2}$ pound meatloaf for supper, but the family only ate $1\frac{1}{2}$ pounds. How much meatloaf was left for sandwiches?

Problem Solving

It is illegal to photocopy this page. Copyright © 2023, Richard W. Fisher

Review Exercises	Speed Drills

1. $\dfrac{1}{3}$ of $\dfrac{4}{5}$ =

2. $\dfrac{3}{2}$ x $\dfrac{3}{5}$ =

3. $\dfrac{2}{5}$
  $-\dfrac{2}{5}$

4. $\dfrac{2}{5}$
  $+\dfrac{2}{5}$

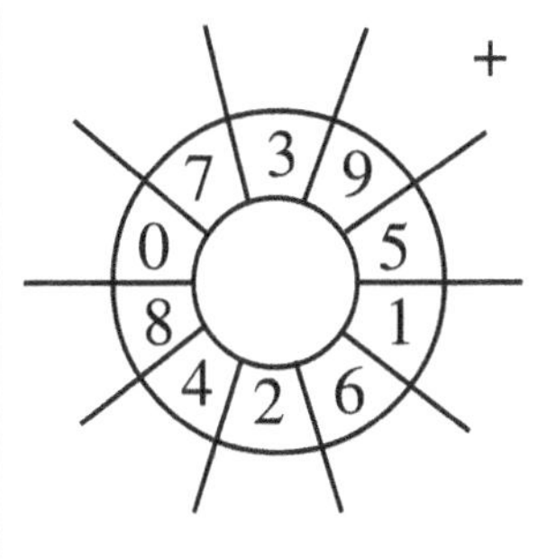

**If the denominator of one fraction and the numerator of another have a common factor, they can be divided out before you multiply the fractions.
Remember, "of" means the same as "x" or "times."**

Examples

4 is a common factor

$$\dfrac{3}{\overset{}{\underset{1}{\cancel{4}}}} \times \dfrac{\overset{2}{\cancel{8}}}{11} = \left(\dfrac{6}{11}\right)$$

3 and 4 are common factors

$$\dfrac{\overset{3}{\cancel{9}}}{\underset{2}{\cancel{8}}} \times \dfrac{\overset{1}{\cancel{4}}}{\underset{1}{\cancel{3}}} = \dfrac{3}{2} = \left(1\dfrac{1}{2}\right)$$

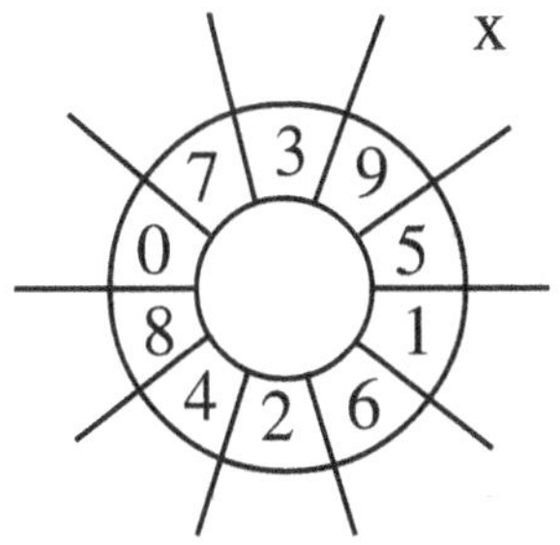

◄ HELPFUL HINTS

S. $\dfrac{3}{5}$ x $\dfrac{5}{7}$ =

S. $\dfrac{9}{10}$ x $\dfrac{5}{3}$ =

1. $\dfrac{2}{5}$ of $\dfrac{3}{4}$ =

2. $\dfrac{2}{6}$ x $\dfrac{3}{5}$ =

3. $\dfrac{15}{16}$ x $\dfrac{3}{5}$ =

4. $\dfrac{3}{4}$ x $\dfrac{7}{9}$ =

5. $\dfrac{4}{3}$ x $\dfrac{6}{7}$ =

6. $\dfrac{5}{6}$ x $\dfrac{7}{10}$ =

7. $\dfrac{3}{4}$ x $\dfrac{3}{5}$ =

8. $\dfrac{2}{7}$ of $\dfrac{14}{15}$ =

9. $\dfrac{8}{9}$ x $\dfrac{3}{4}$ =

10. $\dfrac{1}{6}$ x $\dfrac{4}{5}$ =

1	
2	
3	
4	
5	
6	
7	
8	
9	
10	
Score	

Problem Solving — There are 7 rows of seats in a theater. If each row has 11 seats, how many people can be seated for a show?

It is illegal to photocopy this page. Copyright © 2023, Richard W. Fisher

Speed Drills | Review Exercises

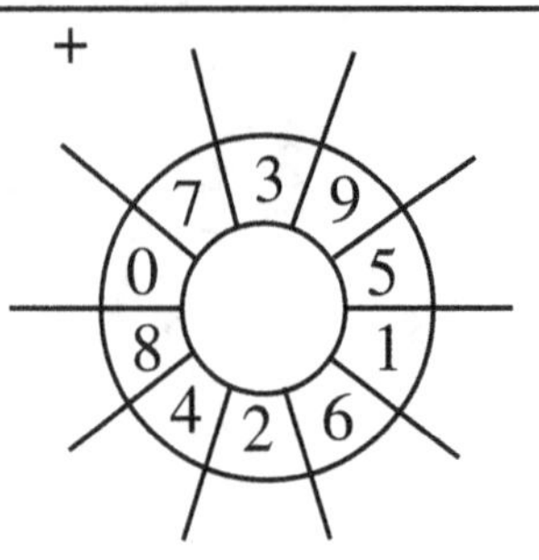

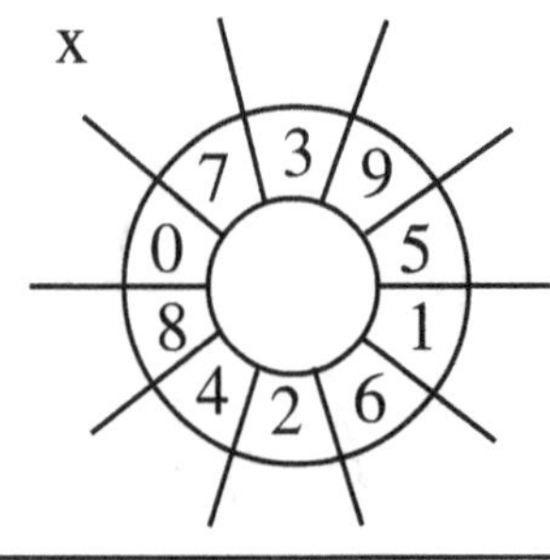

HELPFUL HINTS

1. $\dfrac{2}{3}$ x $\dfrac{5}{6}$ =   2. $\dfrac{1}{3}$ x $\dfrac{2}{5}$ =

3. $\dfrac{2}{5}$
    $+ \dfrac{1}{3}$
    ____

4. $\dfrac{3}{4}$
    $- \dfrac{1}{3}$
    ____

When multiplying whole numbers and fractions, write the whole number as a fraction, check for common factors, then multiply.

Examples:

$\dfrac{2}{3}$ x 15 =

$\dfrac{2}{\cancel{3}_1} \times \dfrac{\cancel{15}^5}{1} = \dfrac{10}{1} = \boxed{10}$

$\dfrac{3}{4}$ x 5 =

$\dfrac{3}{4} \times \dfrac{5}{1} = \dfrac{15}{4} = \boxed{3\dfrac{3}{4}}$

S. $\dfrac{2}{3}$ x 12 =   S. $\dfrac{2}{3}$ x 5 =   1. $\dfrac{3}{4}$ x 8 =

2. 10 x $\dfrac{2}{5}$ =   3. $\dfrac{4}{5}$ x 20 =   4. $\dfrac{2}{7}$ x 4 =

5. $\dfrac{1}{2}$ of 3 =   6. $\dfrac{2}{3}$ x 9 =   7. 6 x $\dfrac{7}{12}$ =

8. $\dfrac{1}{2}$ x 8 =   9. $\dfrac{4}{5}$ of 10 =   10. $\dfrac{2}{3}$ x 4 =

1	
2	
3	
4	
5	
6	
7	
8	
9	
10	
Score	

36   A class has 32 students. If $\dfrac{1}{2}$ of the class is made up of girls, how many girls are there?

Problem Solving

It is illegal to photocopy this page. Copyright © 2023, Richard W. Fisher

	Review Exercises	Speed Drills

1. $2 \overline{)65}$ 2. $\frac{3}{4} \times 16 =$

3. $\frac{2}{3} \times 5 =$ 4. Change $2\frac{1}{2}$ to an improper fraction

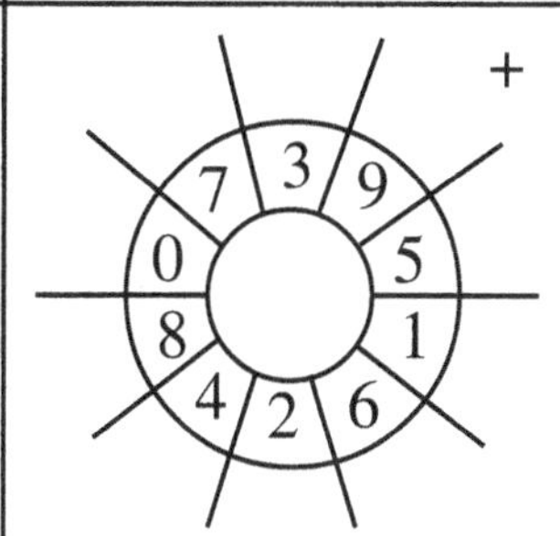

To multiply mixed numerals, first change them to improper fractions, then multiply them.

Example:

$$1\frac{1}{2} \times 1\frac{2}{3} =$$

$$\frac{\cancel{3}^{1}}{2} \times \frac{5}{\cancel{3}_{1}} = \frac{5}{2} = \boxed{2\frac{1}{2}}$$

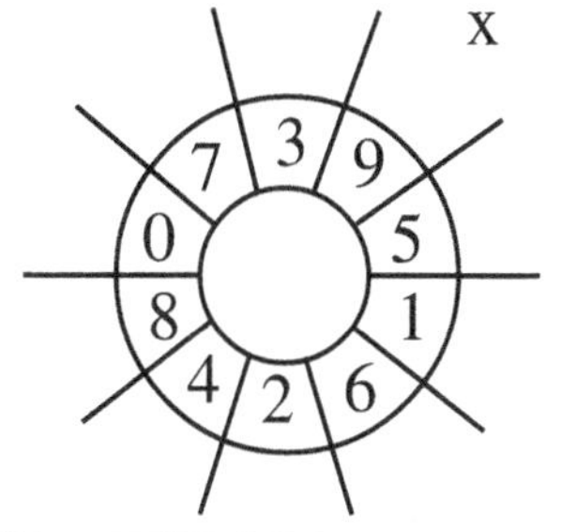

HELPFUL HINTS

S. $\frac{1}{2} \times 2\frac{1}{2} =$ S. $1\frac{1}{4} \times 1\frac{1}{5} =$ 1. $\frac{1}{3} \times 1\frac{1}{2} =$

2. $2\frac{1}{2} \times 2 =$ 3. $4 \times 1\frac{1}{2} =$ 4. $\frac{2}{3} \times 1\frac{1}{2} =$

5. $1\frac{1}{4} \times \frac{3}{10} =$ 6. $1\frac{1}{2} \times 1\frac{1}{3} =$ 7. $\frac{3}{4} \times 2\frac{2}{3} =$

8. $\frac{1}{2} \times 2\frac{3}{7} =$ 9. $2 \times 1\frac{1}{3} =$ 10. $1\frac{1}{4} \times \frac{3}{5} =$

1	
2	
3	
4	
5	
6	
7	
8	
9	
10	
Score	

Problem Solving   A man can run 6 miles per hour. How far can he run in $1\frac{1}{2}$ hours?

37

It is illegal to photocopy this page. Copyright © 2023, Richard W. Fisher

Speed Drills

+

x

HELPFUL HINTS

Review Exercises

1. $\dfrac{1}{5}$
    $+ \dfrac{3}{5}$
    ___

2. $\dfrac{3}{5}$
    $- \dfrac{1}{5}$
    ___

3. $\dfrac{2}{3} \times \dfrac{6}{7} =$

4. $\dfrac{1}{4} \times 8 =$

To find the reciprocal of a common fraction, invert the fraction.

To find the reciprocal of a mixed numeral, first change the mixed number to an improper fraction, then invert it.

To find the reciprocal of a whole number, first make a fraction, then invert it.

Examples

The reciprocal of:

$\dfrac{3}{5}$ is $\dfrac{5}{3}$ or $\left(1\dfrac{2}{3}\right)$

$2\dfrac{1}{2} = \dfrac{5}{2}$ is $\left(\dfrac{2}{5}\right)$

$7 = \dfrac{7}{1}$ is $\left(\dfrac{1}{7}\right)$

1	
2	
3	
4	
5	
6	
7	
8	
9	
10	
Score	

Find the reciprocals of each number:

S. $\dfrac{3}{4}$

S. $1\dfrac{1}{2}$

1. $\dfrac{1}{3}$

2. $\dfrac{7}{8}$

3. $1\dfrac{1}{3}$

4. 12

5. $\dfrac{2}{5}$

6. $\dfrac{1}{7}$

7. 5

8. $\dfrac{2}{3}$

9. $\dfrac{1}{8}$

10. $3\dfrac{1}{2}$

38

Four boys decided to wash cars to earn money. They washed 6 cars and earned 84 dollars. If they divided it equally, how much did each boy get?

Problem Solving

It is illegal to photocopy this page. Copyright © 2023, Richard W. Fisher

Review Exercises	Speed Drills

1.   Find the reciprocal of 5.

2.   Find the reciprocal of $\frac{2}{3}$.

3.   Find the reciprocal of $1\frac{1}{4}$.   4. $\frac{1}{2} \times \frac{2}{3} =$

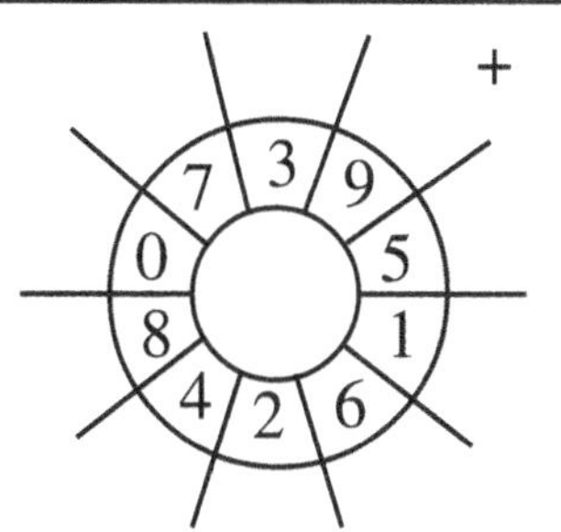

To divide fractions, first find the reciprocal of the second number, then multiply the fractions.

Examples:

$$\frac{2}{3} \div \frac{1}{2} =$$

$$\frac{2}{3} \times \frac{2}{1} = \frac{4}{3} = \boxed{1\frac{1}{3}}$$

$$1\frac{1}{2} \div 2 =$$

$$\frac{3}{2} \times \frac{1}{2} = \boxed{\frac{3}{4}}$$

$$2\frac{1}{2} \div 1\frac{1}{2} =$$

$$\frac{5}{2} \div \frac{3}{2} =$$

$$\frac{5}{\cancel{2}_1} \times \frac{\cancel{2}^1}{3} = \boxed{1\frac{2}{3}}$$

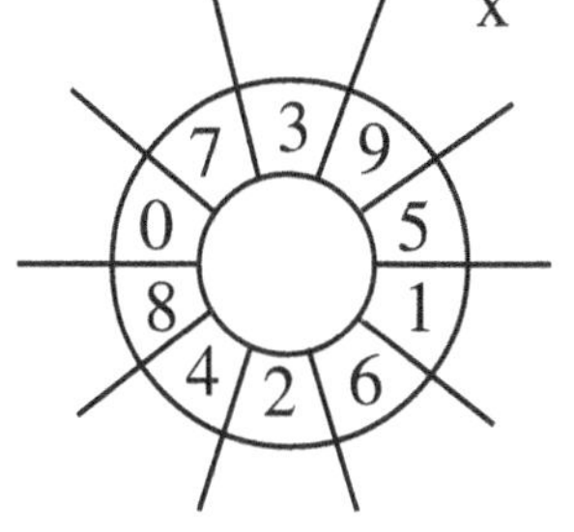

HELPFUL HINTS

S. $\frac{1}{2} \div \frac{1}{3} =$ S. $2\frac{1}{2} \div 2 =$ 1. $\frac{2}{3} \div \frac{1}{2} =$

2. $\frac{3}{5} \div \frac{1}{3} =$ 3. $\frac{2}{5} \div \frac{1}{2} =$ 4. $\frac{2}{5} \div 3 =$

5. $3 \div \frac{1}{2} =$ 6. $1\frac{1}{2} \div \frac{1}{2} =$ 7. $2\frac{1}{2} \div \frac{1}{2} =$

8. $2\frac{1}{6} \div 2 =$ 9. $\frac{2}{7} \div \frac{1}{3} =$ 10. $5 \div \frac{5}{6} =$

1	
2	
3	
4	
5	
6	
7	
8	
9	
10	
Score	

Problem Solving   The coach wants a field which is 36 yards long divided into 3 equal sections for a relay race course. How long will each section be?

39

It is illegal to photocopy this page. Copyright © 2023, Richard W. Fisher

Be sure to reduce your answers.

1.
$$\begin{array}{r} \frac{2}{5} \\ + \ \frac{1}{5} \\ \hline \end{array}$$

2.
$$\begin{array}{r} \frac{7}{8} \\ + \ \frac{2}{8} \\ \hline \end{array}$$

3.
$$\begin{array}{r} \frac{2}{5} \\ + \ \frac{1}{2} \\ \hline \end{array}$$

4.
$$\begin{array}{r} 3\frac{1}{5} \\ + \ 2\frac{4}{5} \\ \hline \end{array}$$

5.
$$\begin{array}{r} 2\frac{1}{2} \\ + \ 3\frac{1}{3} \\ \hline \end{array}$$

6.
$$\begin{array}{r} 2\frac{2}{5} \\ + \ 2\frac{1}{2} \\ \hline \end{array}$$

7.
$$\begin{array}{r} \frac{5}{8} \\ - \ \frac{1}{8} \\ \hline \end{array}$$

8.
$$\begin{array}{r} 7 \\ - \ 2\frac{1}{4} \\ \hline \end{array}$$

9.
$$\begin{array}{r} 3\frac{1}{2} \\ - \ 1\frac{1}{3} \\ \hline \end{array}$$

10.
$$\begin{array}{r} 5\frac{3}{5} \\ - \ 2\frac{1}{2} \\ \hline \end{array}$$

11. $\frac{1}{3} \ \times \ \frac{2}{5} \ =$

12. $\frac{3}{4} \ \times \ \frac{5}{6} \ =$

13. $\frac{3}{4} \ \times \ 8 \ =$

14. $\frac{1}{2} \ \times \ 2\frac{1}{2} \ =$

15. $1\frac{1}{3} \ \times \ 1\frac{1}{4} \ =$

16. $\frac{1}{3} \ \div \ \frac{1}{2} \ =$

17. $\frac{1}{3} \ \div \ \frac{2}{5} \ =$

18. $3 \ \div \ \frac{1}{2} \ =$

19. $2\frac{1}{2} \ \div \ 2 \ =$

20. $1\frac{1}{3} \ \div \ 1\frac{1}{2} \ =$

40

It is illegal to photocopy this page. Copyright © 2023, Richard W. Fisher

	Review Exercises			Speed Drills

1. 23
    16
  + 12

2. $\frac{2}{3}$
  $- \frac{1}{2}$

3. $\frac{3}{4}$
  $+ \frac{1}{3}$

4.   763
   − 432

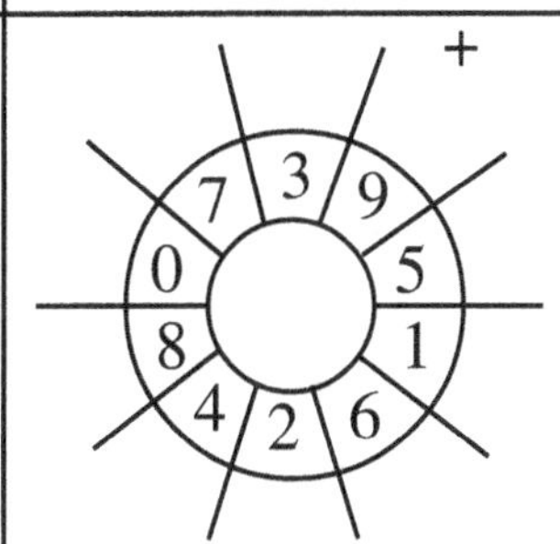

To read decimals, first read the whole number. **Next, read the decimal point as "and."** Then read the number after the decimal point and its place value.

Examples:

3 . 2 = Three **and** two tenths

. 0 0 7 = Seven thousandths

1 4 . 1 6 = Fourteen **and** sixteen hundredths

ones   tens   hundredths   thousandths   ten thousandths   hundred thousandths   millionths

1 . 2 3 4 5 6 7

Express each of the following numbers as words. Write your answer on the line below the number.

S.   2 . 6

S.   12 . 07

1.   . 6

____

____

____

2.   1 . 7

3.   4. 0 0 7

4.   5 . 1 6

____

____

____

5.   1 7 . 0 1 2

6.   . 1 3

7.   4 . 4 2

____

____

____

8.   6 . 0 3

9.   6 . 0 0 3

10. . 0 9

____

____

____

Problem Solving	Maria has 21 trading cards. John has three times as many as Maria has. How many cards does John have?

41

It is illegal to photocopy this page. Copyright © 2023, Richard W. Fisher

Speed Drills	Review Exercises

+

1. 24
    x 23

2. $\frac{2}{3}$ x $\frac{1}{4}$ =

3. $1\frac{1}{2} \div \frac{1}{2}$ =

4. $2 \div \frac{2}{3}$ =

x

HELPFUL HINTS

When reading decimals, remember "and" means decimal point. The fraction part of the decimal ends in "th" or "ths." Be careful about placeholders. Examples:

Four and eight tenths = 4.8

Two and seventeen hundredths = 2.17

Nine thousandths = .009

Write each of the following numbers as decimals. Use the diagram at the bottom of the page if you need help.

S.   Two and six tenths
S.   Three and twelve hundredths
1.   Nine and eight tenths
2.   Two and seventeen hundredths
3.   Thirty-two hundredths
4.   Twenty-two and five tenths
5.   Six thousandths
6.   Two and seven thousandths
7.   Eight and two tenths
8.   Eight and two hundredths
9.   Two and seventeen thousandths
10. Twenty-five hundredths

1	
2	
3	
4	
5	
6	
7	
8	
9	
10	
Score	

ones tenths hundredths thousandths ten thousandths hundred thousandths millionths

1 . 2 3 4 5 6 7

42   If "normal" temperature for a human is $98\frac{3}{5}$ degrees and a man has a temperature of $99\frac{4}{5}$. How much above normal is his temperature? | Problem Solving

It is illegal to photocopy this page. Copyright © 2023, Richard W. Fisher

Review Exercises	Speed Drills

1. $\dfrac{1}{4} \div \dfrac{1}{3} =$ 2. $1\dfrac{1}{2} \div \dfrac{1}{2} =$

3. $\dfrac{3}{4} \div 2 =$ 4. $\dfrac{2}{3} \div 1\dfrac{1}{2} =$

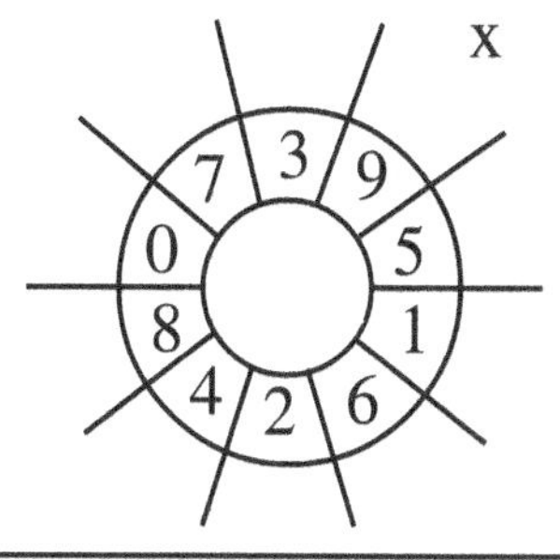

When changing mixed numerals to decimals, remember to place a decimal after the whole number.

Examples: $3\dfrac{3}{10} = \boxed{3.3}$ $3\dfrac{9}{100} = \boxed{3.09}$

$\dfrac{16}{100} = \boxed{.16}$ $\dfrac{7}{1{,}000} = \boxed{.007}$

HELPFUL HINTS

Write each of the following as a decimal. Use the chart at the bottom of the page if you need help.

S. $7\dfrac{7}{10}$ S. $9\dfrac{7}{100}$ 1. $12\dfrac{32}{100}$ 2. $\dfrac{7}{100}$

3. $8\dfrac{9}{10}$ 4. $72\dfrac{9}{100}$ 5. $\dfrac{16}{100}$ 6. $7\dfrac{18}{100}$

7. $6\dfrac{1}{10}$ 8. $4\dfrac{6}{100}$ 9. $12\dfrac{6}{10}$ 10. $7\dfrac{19}{1{,}000}$

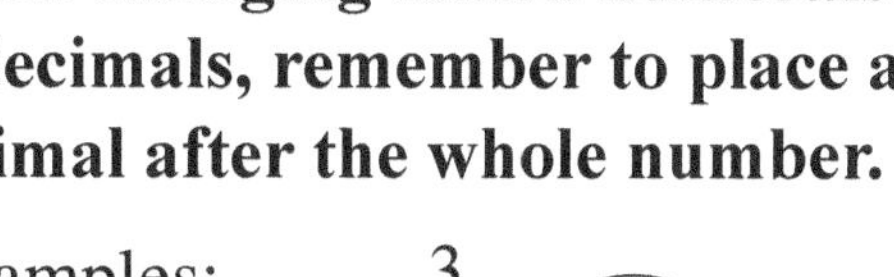

Score	
1	
2	
3	
4	
5	
6	
7	
8	
9	
10	

Problem Solving   One block is $\dfrac{1}{2}$ inch high. How high would a stack of 4 blocks be?

43

It is illegal to photocopy this page. Copyright © 2023, Richard W. Fisher

Speed Drills | Review Exercises

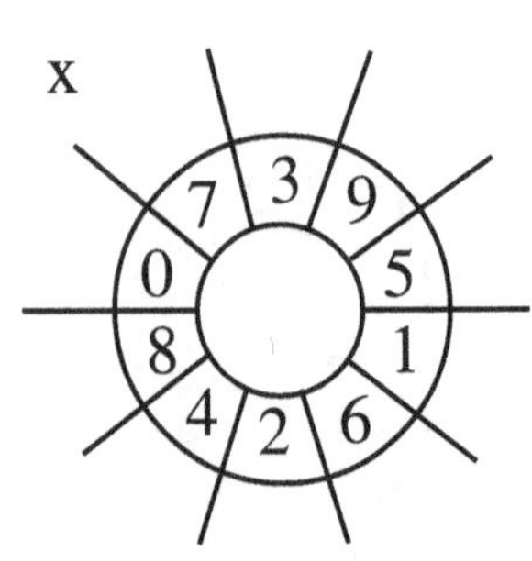

1. $\dfrac{1}{5}$ $+ \dfrac{2}{5}$

2. $5 - 2\dfrac{1}{7}$

3. $\dfrac{4}{5}$ x $\dfrac{3}{8}$ =

4. $\dfrac{1}{2} \div \dfrac{1}{4} =$

Decimals can easily be changed to mixed numerals and fractions. **Remember that the whole number is always to the left of the decimal point.**

Examples:

$$2.6 = \left(2\dfrac{6}{10}\right) \qquad .21 = \left(\dfrac{21}{100}\right)$$

$$3.07 = \left(3\dfrac{7}{100}\right) \qquad 1.012 = \left(1\dfrac{12}{1,000}\right)$$

HELPFUL HINTS

Change each of the following to a mixed numeral or a fraction. Use the chart if you need help.

S. $3.2 =$ S. $5.003 =$ 1. $5.6 =$

2. $.07 =$ 3. $6.09 =$ 4. $7.9 =$

5. $13.015 =$ 6. $.019 =$ 7. $7.008 =$

8. $9.07 =$ 9. $.008 =$ 10. $5.725 =$

	1
	2
	3
	4
	5
	6
	7
	8
	9
	10
	Score

ones   tenths   hundredths   thousandths   ten thousandths   hundred thousandths   millionths

1 . 2   3   4   5   6   7

44

A classroom has 5 rows of chairs, each with 6 chairs in it. Three of the chairs are empty. How many chairs are taken? | **Problem Solving**

It is illegal to photocopy this page. Copyright © 2023, Richard W. Fisher

Review Exercises	Speed Drills

1.   Reduce $\frac{6}{10}$ to its lowest terms.

2.   Change $\frac{7}{2}$ to a mixed numral.

3.   Write 2.7 in words.

4.   Change 3.08 to a mixed numeral.

Zeroes can be put to the right of a decimal without changing its value. This helps when comparing values of decimals.

< means "less than."
> means "greater than."

Examples:

Compare 2.3 and 2.7
2.3 < 2.7

Compare 4.3 and 4.28
4.3 = 4.30 so 4.3 > 4.28

HELPFUL HINTS

Place a < or a > to compare each pair of decimals.

S.   7.6 ☐ 7.3   S.   .2 ☐ .17   1.   6.7 ☐ 5.9

2.   8.3 ☐ 8.9   3.   .4 ☐ .51   4.   .4 ☐ .51

5.   .12 ☐ .3   6.   .7 ☐ .72   7.   3.8 ☐ 3.21

8.   .3 ☐ .005   9.   2.31 ☐ 2.49   10.   6.2 ☐ 6.13

	Score
1	
2	
3	
4	
5	
6	
7	
8	
9	
10	
Score	

Problem Solving   A family is planning a 28-mile hike. If they hike 7 miles per day, how many days will the hike take?

45

It is illegal to photocopy this page. Copyright © 2023, Richard W. Fisher

Speed Drills

+

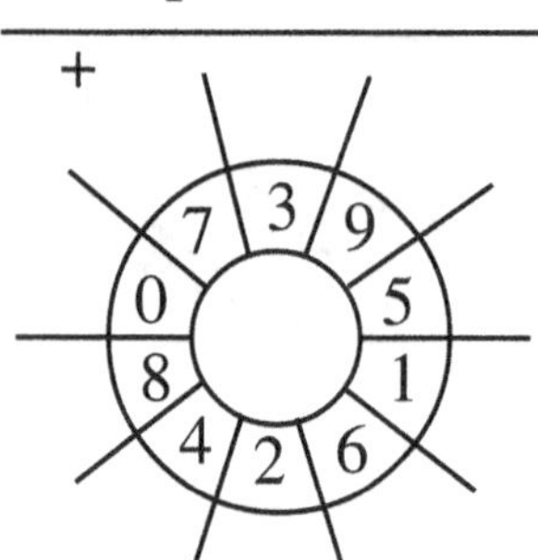

x

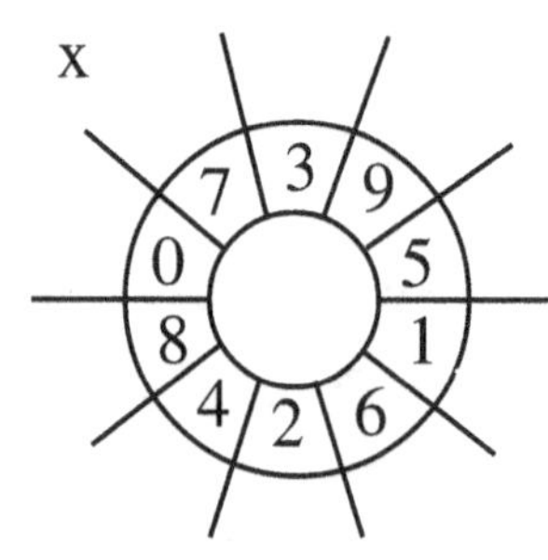

HELPFUL HINTS ▶

Review Exercises

1. $\dfrac{2}{5} \div \dfrac{1}{2} =$

2. $\dfrac{2}{5} \div 2 =$

3. $\begin{array}{r} \dfrac{3}{8} \\ \dfrac{1}{8} \\ + \ \dfrac{1}{8} \end{array}$

4. $\begin{array}{r} 2\dfrac{1}{5} \\ + \ 3\dfrac{3}{5} \end{array}$

To add decimals, first line up the decimal points vertically, then add as you do whole numbers. Be sure you put the decimal point in your answer. Sometimes you need to put zeroes to the right of the decimal point as place holders.

Example:

$3.16 + 2.4 + 6 =$

$$\begin{array}{r} 3.16 \\ 2.40 \\ + \ 6.00 \\ \hline 11.56 \end{array}$$

S. $\begin{array}{r} 3.16 \\ 2.3 \\ + \ 3.26 \\ \hline \end{array}$

S. $2.24 + 3.4 + .23 =$

1. $3.14 + 1.4 =$

2. $5.22 + 3.13 =$

3. $\begin{array}{r} 4.263 \\ + \ 3.23 \\ \hline \end{array}$

4. $.34 + .27 =$

5. $\begin{array}{r} 9.63 \\ + \ 3.7 \\ \hline \end{array}$

6. $3.6 + 7.2 =$

7. $\begin{array}{r} .4 \\ + \ .4 \\ \hline \end{array}$

8. $\begin{array}{r} 3.63 \\ + \ 4.64 \\ \hline \end{array}$

9. $\begin{array}{r} 7.12 \\ + \ 2.14 \\ \hline \end{array}$

10. $3.24 + 3.15 =$

1	
2	
3	
4	
5	
6	
7	
8	
9	
10	
Score	

In March it rained 2.3 inches and in April there was 1.6 inches of rain. What was the total amount of rainfall for the two months?

Problem Solving

46

It is illegal to photocopy this page. Copyright © 2023, Richard W. Fisher

	Review Exercises	Speed Drills

1. 3.4
    2.16
+   3.22

2. $3.6 + 2.3 =$

3. $\dfrac{1}{3}$
    $+ \dfrac{2}{3}$

4. Write $3\dfrac{7}{100}$ as a decimal.

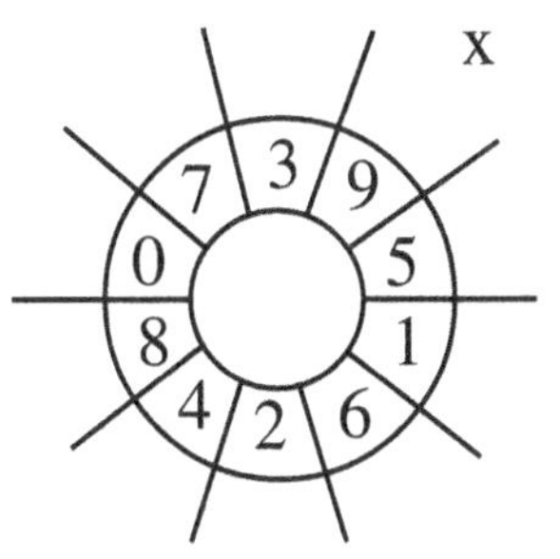

To subtract decimals, first line up the decimal points vertically, then subtract as you do whole numbers. Be sure you put the decimal point in your answer. Sometimes you need to put zeroes to the right of the decimal point as place holders.

Examples:

$3.2 - 1.62 =$

$$\begin{array}{r} {}^{2}\cancel{3}.{}^{11}\cancel{2}\,{}_1 0 \\ -\ 1.62 \\ \hline \overline{(1.58)} \end{array}$$

$7 - 1.63 =$

$$\begin{array}{r} {}^{6}\cancel{7}.{}^{9}\cancel{0}\,{}_1 0 \\ -\ 1.63 \\ \hline \overline{(5.37)} \end{array}$$

HELPFUL HINTS

S. 7.21
  - 3.13

S. $6.8 - 3.14 =$

1. 6.2
  - 3.17

2. 3.26
  - 1.34

3. 5.1
  - 2.43

4. $13.6 - 8.8 =$

5. .7
  - .62

6. 7.43
  - 2.16

7. $3.2 - 1.6 =$

8. .23
  - .124

9. $13.2 - 7.16 =$

10. 7.32
   - 4.25

1	
2	
3	
4	
5	
6	
7	
8	
9	
10	
Score	

Problem Solving | Sue ran a race in 8.6 seconds. Jane ran the race in 8.4 seconds. How much longer did it take Sue to run the race than Jane?

47

It is illegal to photocopy this page. Copyright © 2023, Richard W. Fisher

Speed Drills	Review Exercises

Speed Drills

+

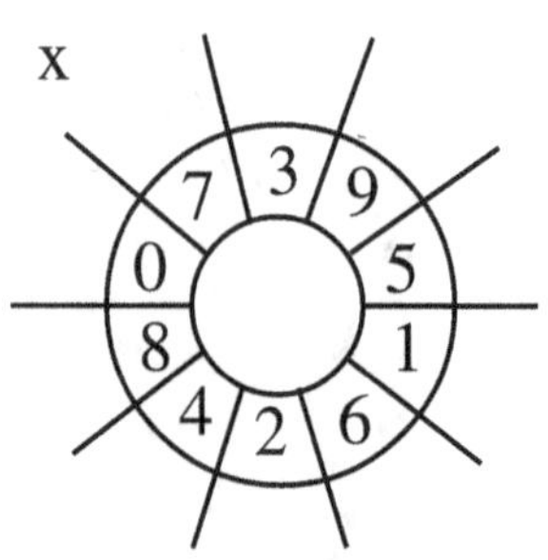

x

HELPFUL HINTS ▶

Review Exercises

1. $\dfrac{1}{3}$
   $-\ \dfrac{1}{3}$

2. $\dfrac{3}{8}$
   $+\ \dfrac{3}{8}$

3. $\dfrac{4}{5}$
   $+\ \dfrac{1}{5}$

4. $\dfrac{1}{3}$
   $+\ \dfrac{2}{3}$
   $\dfrac{1}{3}$

Use what you have learned to solve the following problems.

Remember:
Line up the decimals.
Put the decimal in your answer.
Zeroes may be added to the right of the decimal point.

1	
2	
3	
4	
5	
6	
7	
8	
9	
10	
Score	

S. 3.14
   + 1.32

S. 7.6
   - 2.34

1. 2.45
   + 3.23

2. 3.2
   - 1.4

3. 2.42
   + 3.43

4. 6.2
   - 3.16

5. 7.3
   − 2.6

6. .2
   + .4

7. 7.61
   − 2.43

8. 5.21 + 2.6 + 3.23 =

9. 3.1
   − 1.6

10. 2.43
   + 3.24

48

A boy earned $7.95 mowing the lawn. If he spends $3.34, how much does he have left?

Problem Solving

It is illegal to photocopy this page. Copyright © 2023, Richard W. Fisher

Review Exercises	Speed Drills

1. 24   2. 232   3. 24   4. 124
   x 3   x 4   x 23   x 23

Multiply as you would with whole numbers. Count the number of decimal places and place the decimal point properly in the product.

Examples:

2.32 ← 2 places
x 3
(6.96) ← 2 places

1.4 ← 1 place
x 23
42
280
(3 2 .2) ← 1 place

HELPFUL HINTS

S.   3.43   S.   4.3   1.   2.15
    x 2   x 12   x 3

2.   .423   3.   53.2   4.   8.4
    x 2   x 4   x 2

5.   2.32   6.   2.6   7.   5.14
    x 3   x 5   x 3

8.   43.2   9.   .234   10. 3.62
    x 6   x 4   x 4

1	
2	
3	
4	
5	
6	
7	
8	
9	
10	
Score	

Problem Solving A boy walks 2.4 miles in an hour. If he walks for two hours, how far will he go?

It is illegal to photocopy this page. Copyright © 2023, Richard W. Fisher

Speed Drills	Review Exercises

Speed Drills

+

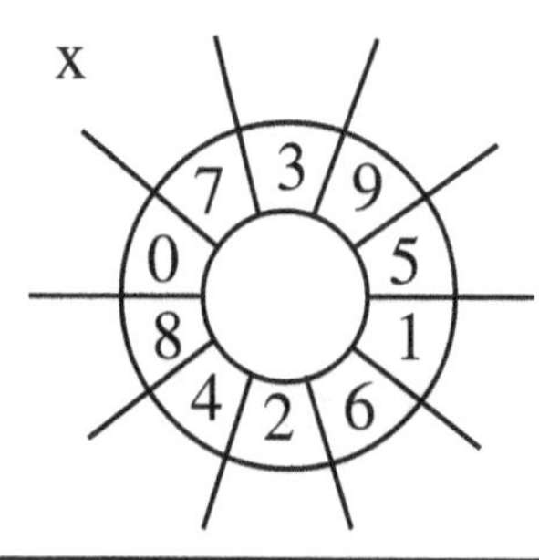

x

HELPFUL HINTS

Review Exercises

1. 2.3
    x 3

2. 4.2
    x 12

3. $\frac{2}{5} \times \frac{1}{4} =$

4. $\frac{1}{2} \div \frac{1}{3} =$

Multiply as you would with whole numbers. Find the total number of decimal places in all numerals, and place the decimal point properly in the product.

Examples:

2.43 ← 2 places
x .2 ← 1 place
.486 ← 3 places

4.3 ← 1 place
x 2.1 ← 1 place
4 3
8 6 0
9.0 3 ← 2 places

S. 3.2
    x .2

S. 4.12
    x 2.3

1. .42
    x .3

2. 6.3
    x .2

3. 21.4
    x .3

4. 2.4
    x 1.3

5. 4.1
    x .3 2

6. 3.42
    x   .2

7. 2.13
    x .5

8. 1.24
    x .6

9. 4.2
    x 3.4

10. 12.4
     x .3

Score

50   A man does laundry 7 days a week. He earns 16 dollars each day. How much does he earn?

Problem Solving

It is illegal to photocopy this page. Copyright © 2023, Richard W. Fisher

Review Exercises	Speed Drills

1. $2\frac{1}{2} \div \frac{1}{2} =$ 2. $2 \times 1\frac{1}{2} =$

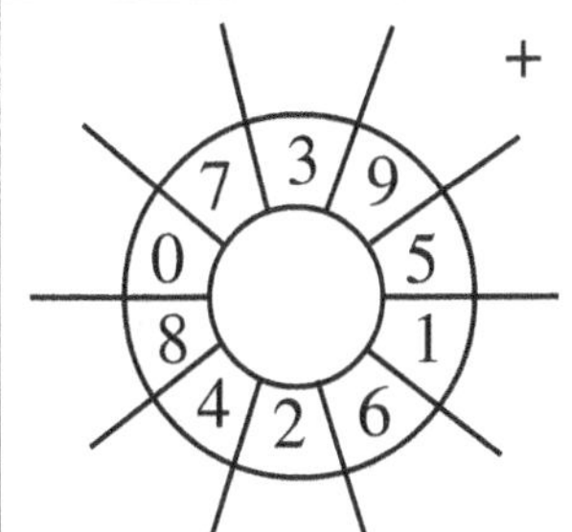

3. $\frac{4}{9} \times \frac{3}{8} =$ 4. $5 \div \frac{1}{2} =$

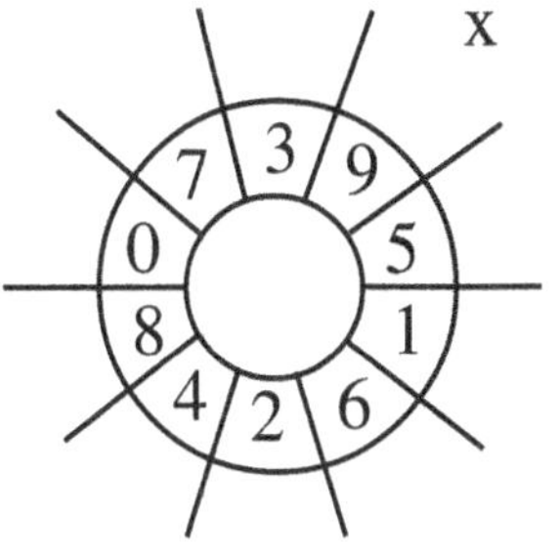

Use what you have learned to solve the following problems.

*** Be careful when placing the decimal in the product.**

HELPFUL HINTS

1	
2	
3	
4	
5	
6	
7	
8	
9	
10	
Score	

S.   2 . 4   S.   4 . 2   1.   1 . 2 3
   x   3   x   5   x   3

2.   3 4 . 3   3.   . 1 4   4.   . 3 4
   x   . 4   x 2 3   x   . 3

5.   3 1 . 2   6.   2 . 4 3   7.   . 4 3 6
   x   6   x   2   x   2

8.   4 . 3 2   9.   2 . 4 3   10.   2.5
   x   2   x   . 5   x   2 3

Problem Solving	A store has shirts on sale for $7.24 each. How much money would John need to buy 3 shirts?

It is illegal to photocopy this page. Copyright © 2023, Richard W. Fisher

Speed Drills

+

x

HELPFUL HINTS

Review Exercises

1. $2\overline{)42}$ 2. $3\overline{)13.2}$

3. $3\overline{)7261}$ 4. $20\overline{)469}$

Divide just as you do with whole numbers. Place the decimal points directly up.

Be careful. Sometimes place holders are necessary.

Examples:

$$\begin{array}{r} 2.3 \\ 3\overline{)6.9} \\ -6\downarrow \\ \hline 9 \\ -9 \\ \hline 0 \end{array}$$

$$\begin{array}{r} .045 \\ 3\overline{).135} \\ -12\downarrow \\ \hline 15 \\ -15 \\ \hline 0 \end{array}$$

1	
2	
3	
4	
5	
6	
7	
8	
9	
10	
Score	

S. $2\overline{)4.6}$ S. $3\overline{)13.2}$ 1. $2\overline{)2.6}$

2. $2\overline{)8.4}$ 3. $4\overline{)8.44}$ 4. $3\overline{)1.53}$

5. $5\overline{)2.55}$ 6. $3\overline{)6.45}$ 7. $4\overline{).928}$

8. $5\overline{)6.15}$ 9. $3\overline{)6.69}$ 10. $2\overline{)13.2}$

52

A man buys four 15-foot boards. He wants to make shelves that are 3 feet long. How many shelves can he make from his boards?

Problem Solving

It is illegal to photocopy this page. Copyright © 2023, Richard W. Fisher

Review Exercises	Speed Drills

Review Exercises

1. $3\overline{)2.4}$ 2. $2\overline{)7.3}$

3. $5\overline{).13}$ 4. $2\overline{).012}$

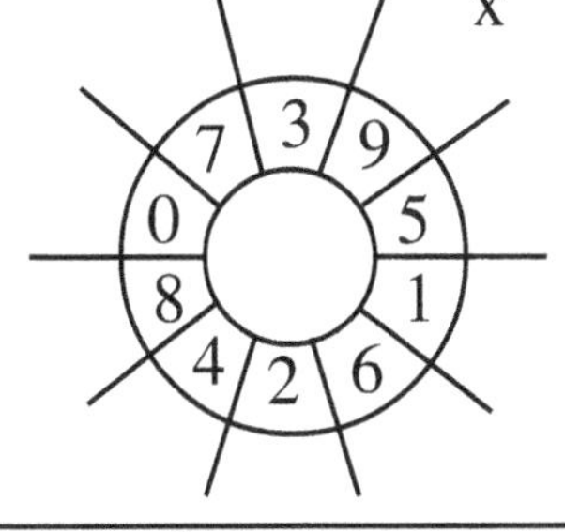

To change a fraction to a decimal, divide the numerator by the denominator. Add as many zeroes as needed.

Examples:

$$\frac{1}{4} = 4\overline{)1.00} \quad \frac{.25}{}$$
$$\frac{-8\downarrow}{20}$$
$$\frac{-20}{0}$$

$$\frac{1}{5} = 5\overline{)1.0} \quad \frac{.2}{}$$
$$\frac{-10}{0}$$

HELPFUL HINTS

Change each of the following fractions to a decimal.

S. $\dfrac{1}{2}$ S. $\dfrac{2}{5}$ 1. $\dfrac{1}{4}$

2. $\dfrac{3}{5}$ 3. $\dfrac{3}{4}$ 4. $\dfrac{2}{5}$

5. $\dfrac{4}{5}$ 6. $\dfrac{1}{5}$ 7. $\dfrac{3}{5}$

8. $\dfrac{9}{10}$ 9. $\dfrac{7}{10}$ 10. $\dfrac{3}{10}$

1	
2	
3	
4	
5	
6	
7	
8	
9	
10	
Score	

Problem Solving | A girl earned 60 dollars by washing cars. She put $\frac{1}{3}$ of her money into the bank. How much did she save?

It is illegal to photocopy this page. Copyright © 2023, Richard W. Fisher

53

Speed Drills	Review Exercises

+

x

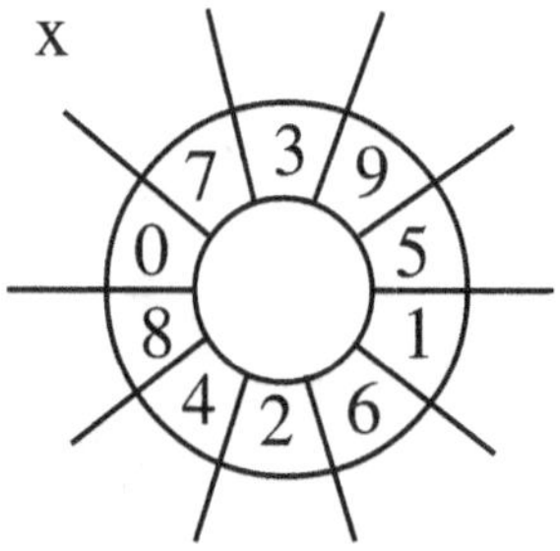

1. $2\overline{)4.8}$
2. $3\overline{).09}$
3. $2\overline{).23}$
4. change $\frac{2}{5}$ to a decimal

Use what you have learned to solve the following problems.

HELPFUL HINTS

*** Add as many zeroes as necessary.**
*** Placeholders may be necessary.**
*** Place decimal points properly.**

	1
	2
	3
	4
	5
	6
	7
	8
	9
	10
	Score

S. $2\overline{)2.4}$

S. $6\overline{).24}$

1. $3\overline{)6.9}$

2. $5\overline{).23}$

3. $5\overline{)3.2}$

4. $3\overline{)2.31}$

5. $5\overline{)2.5}$

6. $2\overline{)2.6}$

7. Change $\frac{4}{5}$ to a decimal.

8. $6\overline{)3.24}$

9. $4\overline{)12.4}$

10. $2\overline{)6.82}$

Joy weighed 120.5 pounds in January. In June, she lost 3.2 pounds. How much did she weigh in June?

Problem Solving

It is illegal to photocopy this page. Copyright © 2023, Richard W. Fisher

1. $\begin{aligned}2.34\\+\ 2.13\end{aligned}$ 2. $3.2 + 1.3 =$

3. $\begin{aligned}2.8\\+\ 6.4\end{aligned}$ 4. $\begin{aligned}7.23\\-\ 2.12\end{aligned}$ 5. $\begin{aligned}5.1\\+\ 2.23\end{aligned}$

6. $7.6 - 2.3 =$ 7. $\begin{aligned}2.3\\\times\ \ 3\end{aligned}$

8. $\begin{aligned}2.43\\\times\ \ 4\end{aligned}$ 9. $\begin{aligned}2.42\\\times\ \ 3\end{aligned}$ 10. $\begin{aligned}42.3\\\times\ \ 2\end{aligned}$

11. $\begin{aligned}.63\\\times\ \ 7\end{aligned}$ 12. $\begin{aligned}3.12\\\times\ \ 3\end{aligned}$ 13. $2\,)\overline{\,2.46\,}$

14. $3\,)\overline{\,3.63\,}$ 15. $2\,)\overline{\,.684\,}$ 16. $5\,)\overline{\,.55\,}$

17. $5\,)\overline{\,3.25\,}$ 18. $3\,)\overline{\,6.6\,}$

19. Change $\frac{1}{5}$ to a decimal. 20. Change $\frac{1}{2}$ to a decimal.

1	
2	
3	
4	
5	
6	
7	
8	
9	
10	
11	
12	
13	
14	
15	
16	
17	
18	
19	
20	

It is illegal to photocopy this page. Copyright © 2023, Richard W. Fisher

Speed Drills	Review Exercises

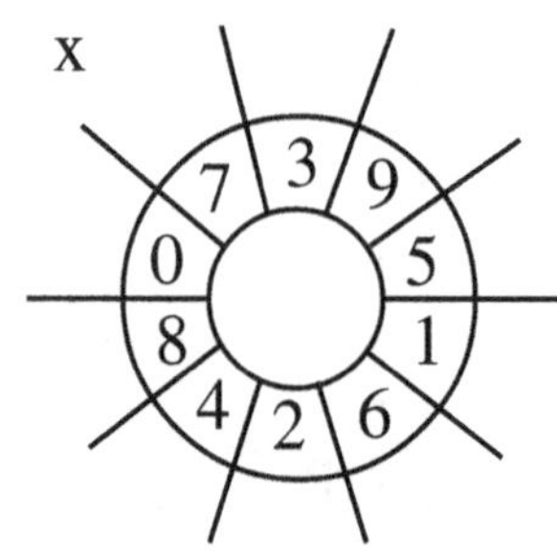

Review Exercises

1. $\dfrac{3}{4} - \dfrac{2}{4}$

2. $\dfrac{2}{5} + \dfrac{2}{5}$

3. $\dfrac{1}{2} \times \dfrac{1}{4} =$

4. $\dfrac{1}{3} \div \dfrac{1}{2} =$

A ratio compares two numbers or groups of objects.

Example: ○ ○ ○   For every 3 circles there are 4 squares.
□ □ □ □

The ratio can be written in the following ways:

3 to 4,   3 : 4,   and $\dfrac{3}{4}$ Each of these is read as "3 to 4."

Ratios are often written in fraction form.

HELPFUL HINTS

Write each of the following ratios as a fraction:

S.   5 nickels to 3 dimes   S.   9 horses to 4 cows

1.   7 to 2   2.   6 children to 5 adults

3.   4 books to 3 pencils   4.   5 bats to 3 balls

5.   6 : 5   6.   8 to 3

7.   7 dimes to 3 pennies   8.   6 chairs to 4 desks

9.   4 cats to 8 dogs   10.   9 : 3

| 1 |
| 2 |
| 3 |
| 4 |
| 5 |
| 6 |
| 7 |
| 8 |
| 9 |
| 10 |
| Score |

56

Mary's car can drive 15 miles with a gallon of gas. If she has 4 gallons in the gas tank, how far can she go before she runs out?

Problem Solving

It is illegal to photocopy this page. Copyright © 2023, Richard W. Fisher

Speed Drills	Review Exercises

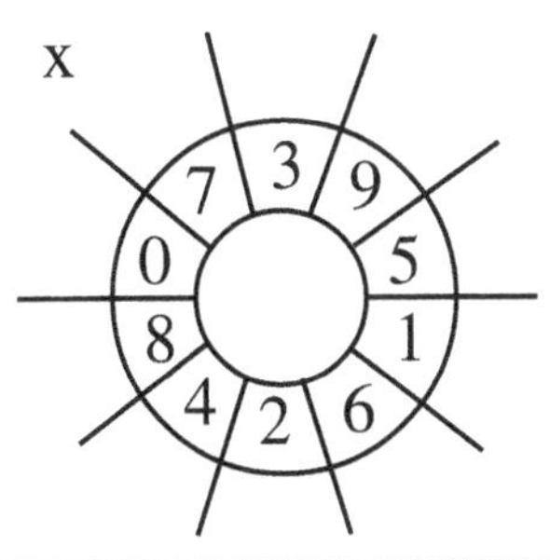

Speed Drills

+ (wheel: 7 3 9 0 5 8 1 4 2 6)

x (wheel: 7 3 9 0 5 8 1 4 2 6)

HELPFUL HINTS

Review Exercises

1. $336 + 12 =$

2. $\begin{array}{r} 73 \\ -\ 14 \end{array}$

3. $\begin{array}{r} 241 \\ \times\ \ 3 \end{array}$

4. $\begin{array}{r} 156 \\ -\ 27 \end{array}$

Percent means "per hundred" or "hundredths." If a fraction is expressed as hundredths, it can easily be written as a percent.

Examples:

$$\frac{7}{100} = \boxed{7\%} \qquad \frac{16}{100} = \boxed{16\%} \qquad \frac{3}{10} = \frac{30}{100} = \boxed{30\%}$$

*** Be sure to change tenths to hundredths.**

1	
2	
3	
4	
5	
6	
7	
8	
9	
10	
Score	

Change each of the following to percents.

S. $\dfrac{12}{100} =$ S. $\dfrac{9}{10} =$ 1. $\dfrac{6}{100} =$ 2. $\dfrac{23}{100} =$

3. $\dfrac{2}{10} =$ 4. $\dfrac{34}{100} =$ 5. $\dfrac{75}{100} =$ 6. $\dfrac{1}{100} =$

7. $\dfrac{7}{10} =$ 8. $\dfrac{15}{100} =$ 9. $\dfrac{80}{100} =$ 10. $\dfrac{62}{100} =$

Problem Solving

A woman bought 4 chairs. If each chair weighed 12.3 pounds, what was the total weight of the chairs?

57

It is illegal to photocopy this page. Copyright © 2023, Richard W. Fisher

Review Exercises

Speed Drills

1. $\dfrac{7}{100}$ = _____%   2. $\dfrac{4}{10}$ = _____%

3. $\dfrac{1}{2}$ x $\dfrac{4}{5}$ =   4. $\begin{array}{r} 7.8 \\ -\ 2.6 \\ \hline \end{array}$

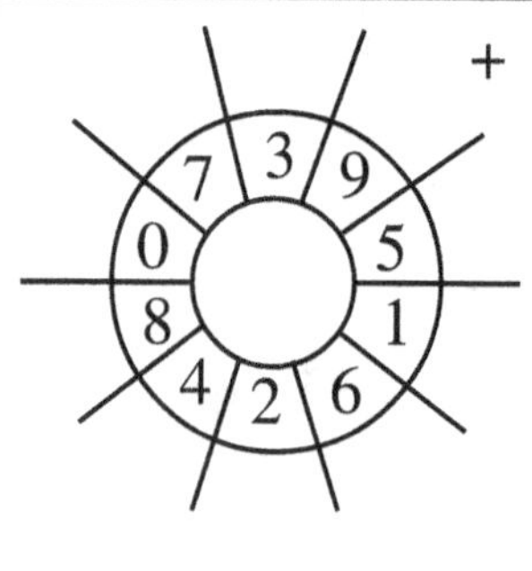

Decimals can easily be changed to percents
   Move the decimal point twice to the right
   and add a percent symbol.
   Be sure to change tenths to hundredths.

"Hundredths" = percent
Example: .15 = 15%   .3 = .30 = 30%

Change each of the following to percents.

S. .12   S. .7   1. .32   2. .02

3. .5   4. .05   5. .6   6. .44

7. .79   8. .4   9. .33   10. .8

1	
2	
3	
4	
5	
6	
7	
8	
9	
10	
Score	

58 There are 4 quarts in a gallon. How many quarts are there in 15 gallons?

Problem
Solving

It is illegal to photocopy this page. Copyright © 2023, Richard W. Fisher

Speed Drills	Review Exercises

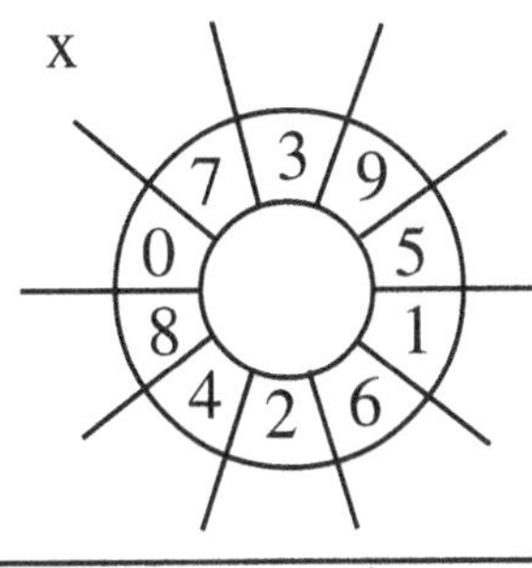

1. Reduce $\dfrac{6}{8}$ to its lowest terms.

2. Change $\dfrac{7}{3}$ to a mixed numeral.

3.
$$\begin{array}{r} 277 \\ + \ 342 \\ \hline \end{array}$$

4.
$$\begin{array}{r} 7\frac{2}{3} \\ - \ 2\frac{1}{3} \\ \hline \end{array}$$

Percents can be easily expressed as decimals and fractions.

HELPFUL HINTS

Examples:

$$25\% = .25 = \frac{25}{100}$$

$$8\% = .08 = \frac{8}{100}$$

$$30\% = .30 = \frac{30}{100} = \frac{3}{10}$$

1	
2	
3	
4	
5	
6	
7	
8	
9	
10	
Score	

Change each percent to a decimal and a fraction.

S. $12\% = .\quad = \text{—}$ S. $4\% = .\quad = \text{—}$ 1. $16\% = .\quad = \text{—}$

2. $6\% = .\quad = \text{—}$ 3. $75\% = .\quad = \text{—}$ 4. $40\% = .\quad = \text{—}$

5. $1\% = .\quad = \text{—}$ 6. $45\% = .\quad = \text{—}$ 7. $12\% = .\quad = \text{—}$

8. $5\% = .\quad = \text{—}$ 9. $50\% = .\quad = \text{—}$ 10. $13\% = .\quad = \text{—}$

Problem Solving

There are 25 students in a class. If 3 are absent, how many are present?

59

It is illegal to photocopy this page. Copyright © 2023, Richard W. Fisher

Review Exercises	Speed Drills

1. $2 \overline{)\, 4.8}$ 2. Change .25 to a percent.

3.   2.35
    x   3

4.   4.25
    x   3

To find the percent of a number, change the percent to a decimal and multiply.

Examples:

Find 3% of 60
.03 x 60

$$\begin{array}{r} 60 \\ \times\ .03 \\ \hline 180 \\ 00 \\ \hline 1.80 \end{array}$$

Find 40% of 35
.4 x 35

$$\begin{array}{r} 35 \\ \times\ .4 \\ \hline 14.0 \end{array}$$

HELPFUL HINTS

S. Find 2% of 60.   S. Find 30% of 40.   1. Find 3% of 52.

2. Find 40% of 35.   3. Find 4% of 30.   4. Find 40% of 30.

5. Find 12% of 40.   6. Find 50% of 16.   7. Find 30% of 70.

8. Find 12% of 50.   9. Find 20% of 80.   10. Find 2% of 40.

Speed Drills	
1	
2	
3	
4	
5	
6	
7	
8	
9	
10	
Score	

Problem Solving

60   A gasoline tank holds $5\frac{1}{2}$ gallons. If $2\frac{1}{2}$ gallons have been used, then how many gallons are left in the tank?

It is illegal to photocopy this page. Copyright © 2023, Richard W. Fisher

Speed Drills

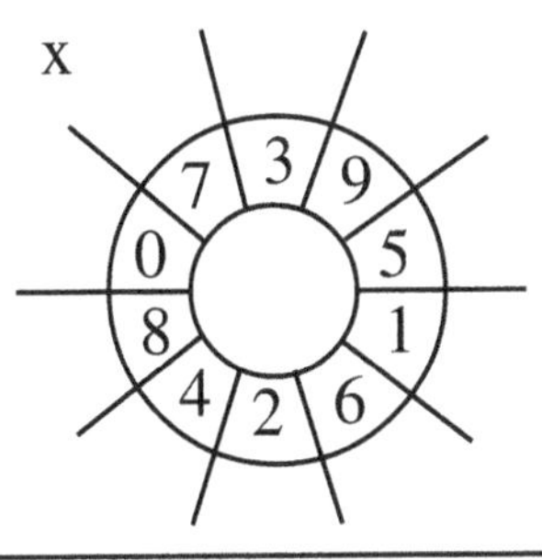

HELPFUL HINTS

Review Exercises

1. Find 13% of 24.

2. Change $\frac{4}{5}$ to a decimal

3. Change 3% to a decimal.

4. Find 20% of 60.

When finding the percent of a number in a word problem, you can change the percent to a decimal. Always express your answer in a short phrase or sentence.

Example:

A team played 60 games and won 75% of them. How many games did they win?

Find 75% of 60
.75 x 60

$$\begin{array}{r} 60 \\ \times\ .75 \\ \hline 300 \\ 420 \\ \hline 45.00 \end{array}$$

Answer: The team won 45 games.

S. George took a test with 20 problems. If he got 15% of the problems correct, how many problems did he get correct?

S. If 6% of the 500 students enrolled in a school are absent, then how many students are absent?

1. A worker earned 80 dollars and put 30% of it into the bank. How many dollars did he put into the bank?

2. A car costs $4,000. If Mr. Smith has saved 20% of this amount, how much did he save?

3. Steve took a test with 30 problems. If he go 30% of the problems correct, how many problems did he get correct?

4. A family's monthly income is $3,000. If 20% of this amount is spent on food, how many dollars are spent on food?

5. There are 40 students in a class. If 60% of the class are boys, then how many boys are in the class?

6. A house that costs $80,000 requires a 20% down payment. How many dollars are required for the down payment?

7. If a car costs $6,000 and loses 30% of its value in one year, how much value will the car lose in one year?

8. A coat is priced $50. If the sales tax is 7% of the price, how much is the sales tax?

	Score
Problem Solving	

A train traveled 83.5 miles per hour. At this rate, how far would it travel in 2 hours?

61

It is illegal to photocopy this page. Copyright © 2023, Richard W. Fisher

	Review Exercises	Speed Drills

1. Find 5% of 60.

2.   43
     x   32

3.   24 + 143 + 25 =

4.   705
      - 324

To change a fraction to a percent first change the fraction to a decimal, then change the decimal to a percent. Move the decimal twice to the right and add a percent symbol.

Examples: $\frac{2}{5}$ 5) 2.0 $\frac{.40}{}$ = 40% $\frac{2}{8} = \frac{1}{4}$ 4) 1.00 .25 = 25%

$\frac{2}{5}$: 5) 2.0 ; − 20 ; 0

$\frac{2}{8} = \frac{1}{4}$: 4) 1.00 ; − 8 ; 20 ; − 20 ; 0

*Sometimes the fraction can be reduced

HELPFUL HINTS

Change each of the following to percents.

S. $\frac{1}{2}$ =

S. $\frac{6}{10}$ =

1. $\frac{4}{20}$ =

2. $\frac{4}{10}$ =

3. $\frac{3}{12}$ =

4. $\frac{5}{10}$ =

5. $\frac{3}{10}$ =

6. $\frac{15}{20}$ =

7. $\frac{6}{12}$ =

8. $\frac{4}{16}$ =

9. $\frac{2}{4}$ =

10. $\frac{3}{15}$ =

1	
2	
3	
4	
5	
6	
7	
8	
9	
10	
Score	

John earned 30 dollars. If he put 20% of it into the bank, how much did he put into the bank?

Problem Solving

It is illegal to photocopy this page. Copyright © 2023, Richard W. Fisher

Speed Drills | Review Exercises

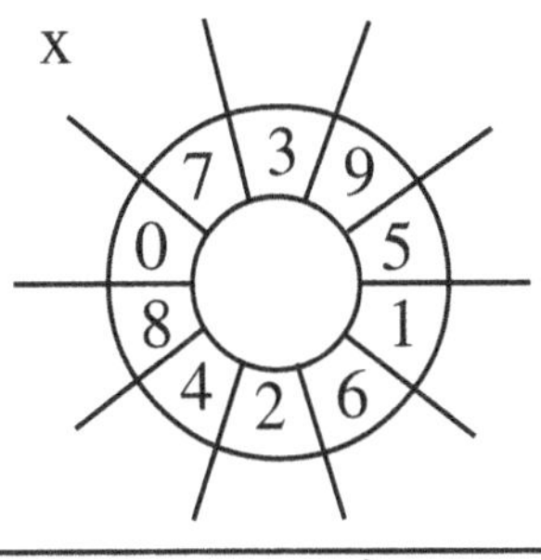

1.   7.12
    x   3

2.   7.6
   - 2.32

3.   2.64
    + 3.37

4. $5\overline{)\,.15}$

When finding the percent, first write a fraction, change the fraction to a decimal, then change the decimal to a percent.

Examples:

3 is what % of 5?

$$\frac{3}{5} \quad 5\overline{)\,3.0} \quad \frac{.60}{} = 60\%$$
$$-\,3\,0$$
$$0$$

2 is what % of 8?

$$\frac{2}{8} = \frac{1}{4} \quad 4\overline{)\,1.00} \quad .25 = 25\%$$
$$-\quad 8\downarrow$$
$$20$$
$$-\quad 20$$
$$0$$

HELPFUL HINTS

	1
	2
	3
	4
	5
	6
	7
	8
	9
	10
Score	

S. 2 is what % of 5?

S. 2 is what % of 10?

1. 4 is what % of 5?

2. 3 is what % of 6?

3. 8 is what % of 10?

4. 4 is what % of 16?

5. 5 is what % of 10?

6. 3 is what % of 15?

7. 1 is what % of 5?

8. 3 is what % of 12?

9. 2 is what % of 4?

10. 5 is what % of 25?

Problem Solving

A man had 75 dollars. One day he spent 12 dollars. The next day he earned 8 dollars. How much money does he have now?

63

It is illegal to photocopy this page. Copyright © 2023, Richard W. Fisher

Review Exercises	Speed Drills

1. Find 12% of 42.

2. Change $\frac{2}{5}$ to a decimal

3. Find 30% of 50.

4. $3 \overline{)\ .18}$

When finding the percent, first write a fraction, change the fraction to a decimal, then change the decimal to a percent.

Example:

A team played 8 games and won 2 of them. What percent of the games did they win?

2 is what % of 8?

$$\frac{2}{8} = \frac{1}{4}$$

$$4 \overline{)\ 1.00} \quad .25 = 25\%$$

They won 25% of the games.

HELPFUL HINTS

S. A test had 15 questions. If Sam got 9 questions correct, what percent did he get correct?

S. In a class of 15 students, 3 are girls. What percent of the class is girls?

1. On a spelling test with 12 words, Susan got 3 correct. What percent of the words did she get correct?

2. A worker earned 20 dollars. If she put 5 dollars into a savings account, what percent of her earnings did she put into a savings account?

3. A team played 5 games and won 3 of them. What percent did they win?

4. A quarterback threw 25 passes and 20 were caught. What percent of the passes were caught?

5. $\frac{2}{5}$ of a class was present at school. What percent of the class was present?

6. A class has an enrollment of 20 students. If 2 are absent, what percent are absent?

7. A team played 12 games and lost 3 games. What percent of the games played did it lose?

8. A class has 20 students. If 8 of them are boys, what percent are boys?

9. On a math test with 10 questions, Jill got 9 of them correct. What percent did she get correct?

10. A pitcher threw 8 pitches. If 4 of them were strikes, what percent were strikes?

1	
2	
3	
4	
5	
6	
7	
8	
9	
10	
Score	

There are 20 questions on a test. If a student got 80% of them correct, how many questions did he get correct?

Problem Solving

64

It is illegal to photocopy this page. Copyright © 2023, Richard W. Fisher

Speed Drills

+

(speed drill wheel: 7, 3, 9, 0, 5, 8, 1, 4, 2, 6)

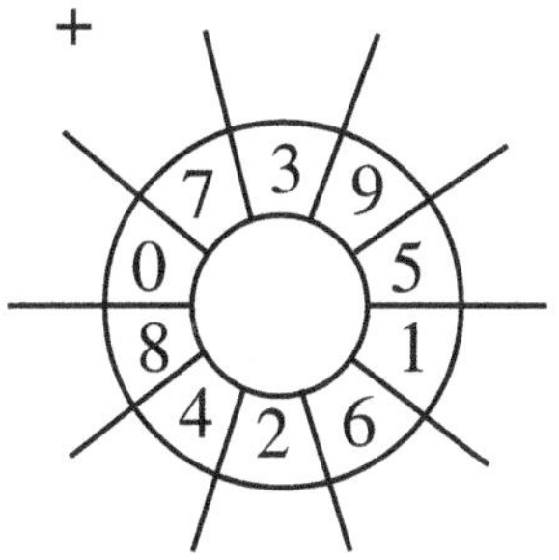

x

(speed drill wheel: 7, 3, 9, 0, 5, 8, 1, 4, 2, 6)

HELPFUL HINTS

1	
2	
3	
4	
5	
6	
7	
8	
9	
10	
Score	

Problem Solving

Review Exercises

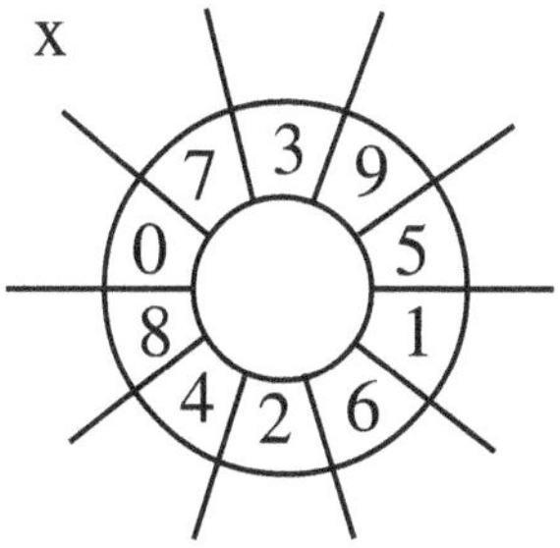

1. Change $\dfrac{7}{100}$ to a percent.   2. Change $\dfrac{9}{10}$ to a percent.

3. Change .03 to a percent.   4. Change .7 to a percent.

Use what you have learned to solve the problems.

Examples:

Find 12% of 60
.12 x 60

$$\begin{array}{r} 60 \\ \times\ .12 \\ \hline 120 \\ 60 \\ \hline (7.20) \end{array}$$

3 is what % of 12?

$\dfrac{3}{12} = \dfrac{1}{4}$

$$4\overline{)1.00}\quad .25 = (25\%)$$
$$\begin{array}{r} -\ 8\downarrow \\ \hline 20 \\ -\ 20 \\ \hline 0 \end{array}$$

S. Find 20% of 30.

S. 2 is what % of 8?

1. Find 3% of 60.

2. Find 12% of 65.

3. 3 is what % of 5?

4. 4 is what % of 8?

5. Find 30% of 60

6. Find 5% of 16

7. 5 is what % of 25?

8. Find 15% of 40.

9. Find 50% of 30.

10. 4 is what % of 5?

There are 30 problems on a test. If a student got 80% of them correct, how many problems did he get correct?

65

It is illegal to photocopy this page. Copyright © 2023, Richard W. Fisher

Review Exercises	Speed Drills

1. $\dfrac{5}{6} - \dfrac{1}{6}$

2. $\dfrac{3}{8} + \dfrac{1}{8}$

3. $\dfrac{3}{5} \times \dfrac{10}{11} =$

4. $\dfrac{2}{3} \div \dfrac{1}{3} =$

Use what you have learned to solve the problem. Examples:

A farmer has 20 cows. If he sells 40% of them, how many does he sell?

Find 40% of 20

.4 x 20

$$\begin{array}{r} 20 \\ \times\ .4 \\ \hline 8.0 \end{array}$$

He sold 8 cows.

In a class of 20 students, 5 are girls. What percent are girls?

5 is what % of 20?

$$\frac{5}{20} = \frac{1}{4}$$

$$\begin{array}{r} 25 = 25\% \\ 4\,\overline{)\,1.00} \\ -\ \ 8\!\downarrow \\ \hline 20 \\ -\ \ 20 \\ \hline 0 \end{array}$$

25% are girls.

HELPFUL HINTS

S. A test has 40 problems. A student got 30% of them correct. How many problems did he get correct?

S. Sue has finished 3 problems on a test. If there are 12 problems on the test, what percent has she finished?

1. A ranch has 500 acres of land. If 60% of the land is used for grazing, then how many acres are used for grazing?

2. A player took 5 shots. If he made 3 of them, what percent did he make?

3. A man earned 20 dollars and spent 60% of it. How much did he spend?

4. A test has 20 questions. If Jane got 15 correct, what percent did she get correct?

5. 4 is what percent of 5?

6. Find 12% of 40.

7. There are 400 students in a school. If 60% eat cafeteria food, how many students eat cafeteria food?

8. A baseball team played 5 games and won 3. What percent did they win?

9. A car costs $6,000. If a down payment of 20% is required, how much is the down payment?

10. 10 players tried out for a team. If only 6 made the team, what percent made the team?

1	
2	
3	
4	
5	
6	
7	
8	
9	
10	
Score	

A man can run 6 miles per hour. At this rate how far can he run in 4 hours?

Problem Solving

66

It is illegal to photocopy this page. Copyright © 2023, Richard W. Fisher

Write numbers 1 and 2 as a ratio expressed in fraction form.

1.   7 nickels to 2 dimes   2.   9 to 4

For numbers 3 and 4 solve each proportion.

3. $\dfrac{4}{5} = \dfrac{?}{10}$ 4. $\dfrac{3}{4} = \dfrac{9}{?}$

Change numbers 5 through 8 to a percent.

5. $\dfrac{17}{100} =$ 6. $\dfrac{7}{10} =$ 7.   .19 =   8.   .6 =

Change numbers 9 through 11 to a decimal and a fraction.

9.   6% = .   =   –   10.   15% = .   =   –

11.   80% = .   =   –

12.   Find 4% of 60.   13.   Find 30% of 40.

14.   6 is what % of 8?   15.   3 is what % of 5?

16.   Change $\dfrac{1}{5}$ to a percent.   17.   Change $\dfrac{1}{4}$ to a percent.

18.   A man earned 40 dollars. If he put 20% of it into the bank, how much did he put into the bank?

19.   A woman baked 8 cakes. If she sold 6 of them, what percent did she sell?

20.   On a test with 20 questions, a student got 70% correct. How many questions did the student get correct?

67

It is illegal to photocopy this page. Copyright © 2023, Richard W. Fisher

Review Exercises	Speed Drills

1. $3.2 + .54 + 2.8 =$ 2. $7.6 - 5.8 =$

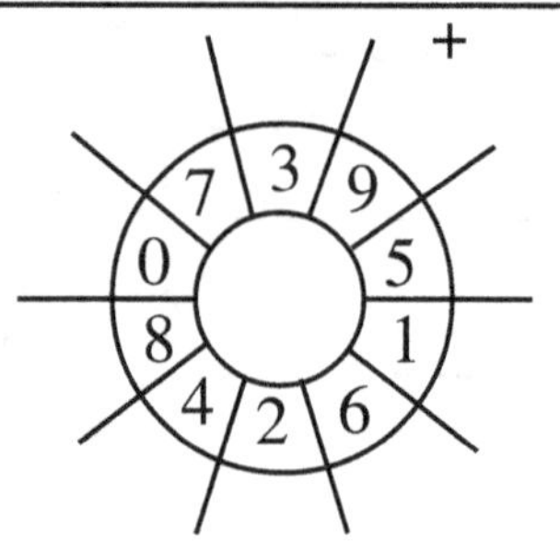

3.   5.6   4. $2 \overline{)9.2}$
       x   2

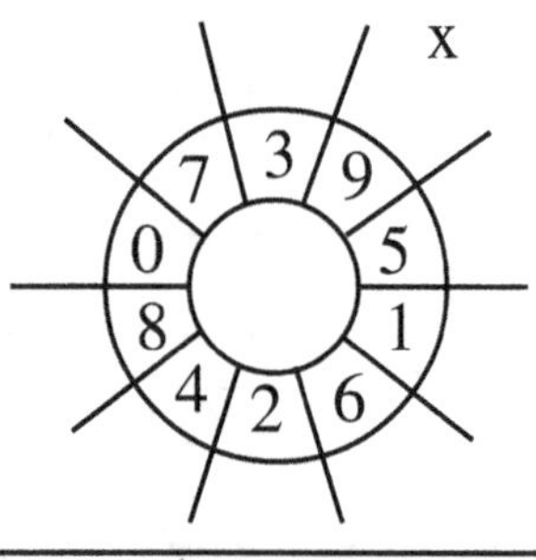

Geometric term:	Point	Line	Plane	Line Segment	Ray
Example:	• P	A B $\longleftrightarrow$	•A •B •C	A B	A B
Symbol:	P	$\overleftrightarrow{AB}$	plane ABC	$\overline{AB}$	$\overrightarrow{AB}$

Use the figure to answer the following.
S. Name 2 points   S. Name 2 line segments

1. Name 2 lines   2. Name 2 rays

3. Name 2 points on $\overleftrightarrow{FD}$

4. Give another name for $\overleftrightarrow{AB}$

5. Give another name for $\overleftrightarrow{ED}$

6. Give another name for $\overleftrightarrow{AC}$

7. Name a line segment on $\overleftrightarrow{FD}$

8. Name 2 rays on $\overleftrightarrow{FE}$

9. Name a ray on $\overleftrightarrow{AC}$

10. What point is common to lines $\overleftrightarrow{FD}$ and $\overleftrightarrow{BE}$?

1	
2	
3	
4	
5	
6	
7	
8	
9	
10	
Score	

68   If a factory can build an engine in 2 hours, how long
will it take to build 10 engines?

Problem Solving

It is illegal to photocopy this page. Copyright © 2023, Richard W. Fisher

Speed Drills

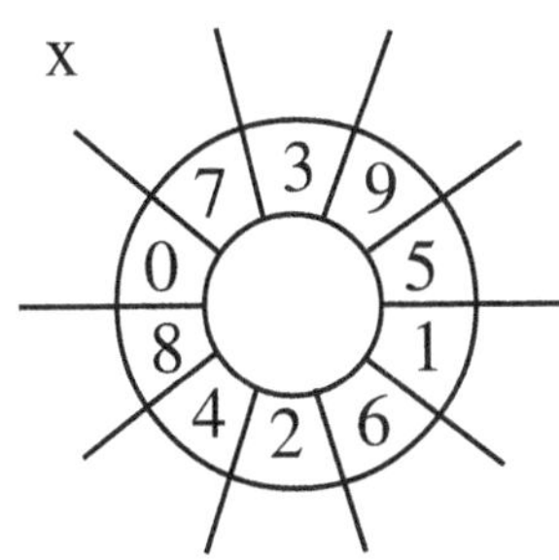

Review Exercises

1. $\dfrac{3}{5}$
$+\ \dfrac{4}{5}$

2. $\dfrac{3}{4}$
$-\ \dfrac{1}{4}$

3. $2 \times \dfrac{3}{4} =$

4. $\dfrac{3}{4} \div \dfrac{1}{4} =$

HELPFUL HINTS

Geometric term:	Parallel Lines	Intersecting Lines	Perpendiculat Lines	Angle	Symbols

Example:

Symbols: $\angle$ DAC, $\angle$ CAD, $\angle$ A   Vertex

Use the figure to answer the following.

S. Name 2 parallel lines

S. Name 2 perpendicular lines

1. Name a pair of intersecting lines

2. Name 2 angles

3. Name 2 angles that have B as their vertex

4. Name 2 angles that have H as their vertex

5. Name 2 lines   6. Name 2 line segments

7. Name 2 rays   8. Name a line segment on $\overleftrightarrow{BH}$

9. Name 2 lines that include point B

10. Give another name for $\angle$ JHI

	1
	2
	3
	4
	5
	6
	7
	8
	9
	10
Score	

Problem Solving   Diane has a piece of cloth 24 yards long. How many pieces 3 yards long can she cut from it?

69

It is illegal to photocopy this page. Copyright © 2023, Richard W. Fisher

Review Exercises	Speed Drills

1. 6 is what% of 8?

2. Find 12% of 80.

3. A man had 300 cows and decided to sell 20% of them. How many cows did he sell?

4. Sue took a test with 20 problems. If she got 12 of the problems correct, then what percent of the problems did she get correct?

right angle

measures 90°

acute angle

measures less than 90°

obtuse angle

measures more than 90°

straight angle

measures 180°

HELPFUL HINTS

When naming an angle, the letter in the middle is the vertex.

Use the figure to answer the following.

S. Name 2 right angles

S. Name 2 acute angles

1. Name 2 obtuse angles

2. Name 2 straight angles

3. What kind of angle is $\angle$ IJG?

4. What kind of angle is $\angle$ EDB?

5. What kind of angle is $\angle$ GBD?

6. What kind of angle is $\angle$ GJK?

7. Name an acute angle which has J as its vertex.

8. Name an obtuse angle that has D as its vertex.

9. Name a right angle which has B as its vertex.

10. Name a straight angle which has D as its vertex.

Speed Drills	
1	
2	
3	
4	
5	
6	
7	
8	
9	
10	
Score	

70

A rope is 35 meters long. If a man wants to cut it into 5 pieces of equal length, how long will each piece be?

Problem Solving

It is illegal to photocopy this page. Copyright © 2023, Richard W. Fisher

Speed Drills	Review Exercises

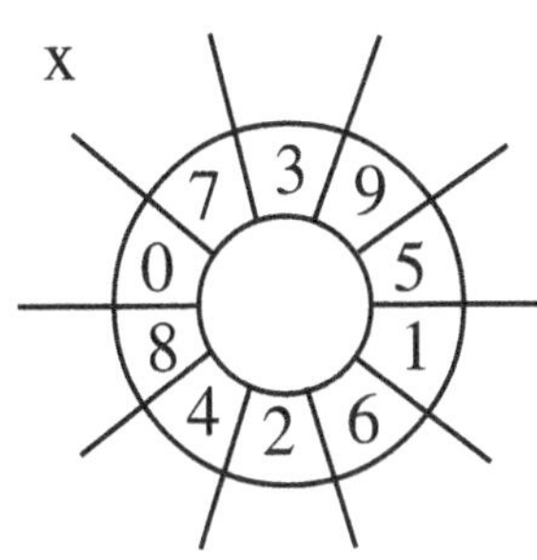

+

1. Change $\dfrac{9}{7}$ to a mixed numeral.

2. Change $2\dfrac{1}{2}$ to an improper fraction

x

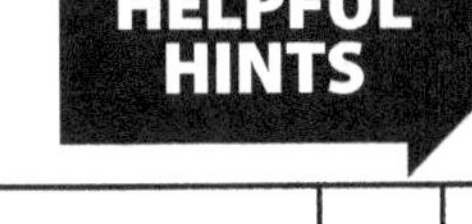

3. Express $\dfrac{8}{10}$ in its lowest terms.

4. $\dfrac{1}{2}$ of $12 =$

HELPFUL HINTS

To use a protractor, follow these rules:
1. **Place the center point of the protractor on the vertex.**
2. **Place the zero mark on one edge of the angle.**
3. **Read the number where the other side of the angle crosses the protractor.**
4. **If the angle is acute, use the smaller number. If the angle is obtuse, use the larger number.**

1
2
3
4
5
6
7
8
9
10
Score

Use the figure to answer the questions. Classify the angle as **right, acute, obtuse,** or **straight**. Then tell how many degrees the angle measures.

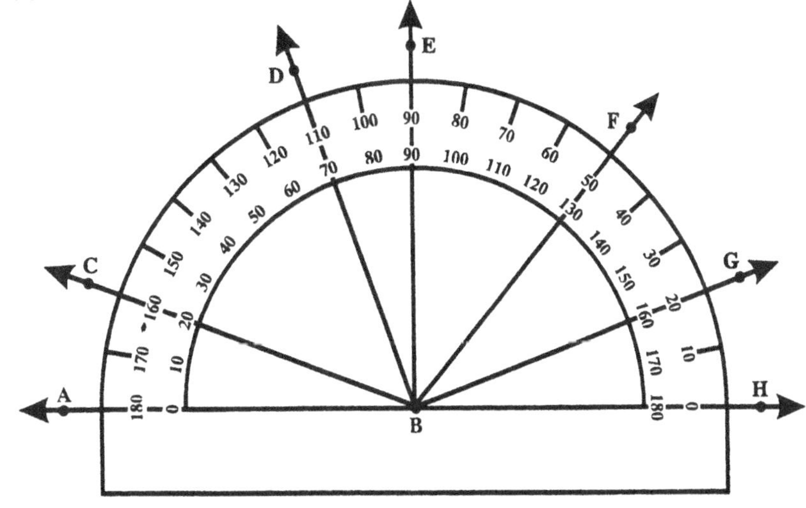

Remember! When naming an angle, the letter in the middle is the vertex.
Example: In $\triangle$ JMI, M is the vertex.

S. $\triangle$HBG   S. $\triangle$DBH   1. $\triangle$EBH   2. $\triangle$CBH
3. $\triangle$GBH   4. $\triangle$DBA   5. $\triangle$ABF   6. $\triangle$FBH
7. $\triangle$ABH   8. $\triangle$ABG   9. $\triangle$EBA   10. $\triangle$FBA

Problem Solving	50 students took a test and 80% received a passing grade. How many students received a passing grade?

71

It is illegal to photocopy this page. Copyright © 2023, Richard W. Fisher

Review Exercises	Speed Drills

1. Change $\frac{4}{5}$ to a percent.   2. Change .9 to a percent.

3. Find 4% of 65.   4. 6 is what percent of 8?

To use a protractor, follow these rules:
1. **Place the center point of the protractor on the vertex.**
2. **Place the zero mark on one edge of the angle.**
3. **Read the number where the other side of the angle crosses the protractor.**
4. **If the angle is acute, use the smaller number. If the angle is obtuse, use the larger number.**

HELPFUL HINTS

Classify each angle as acute, right, obtuse, or straight.

Remember! When naming an angle, the letter in the middle is the vertex.
Example: In △JMI, M is the vertex.

S. △AMC   S. △EMA   1. △DMA   2. △FMJ   3. △FMA
4. △DMJ   5. △EMA   6. △CMA   7. △HMJ   8. △GMJ
9. △IMA 10. △IMJ

1	
2	
3	
4	
5	
6	
7	
8	
9	
10	
Score	

Problem Solving
There are 90 seats in an auditorium. If 83 of the seats are taken, how many seats are empty?

72

It is illegal to photocopy this page. Copyright © 2023, Richard W. Fisher

Speed Drills	Review Exercises

Speed Drills

+

7 3 9
0 5
8 1
4 2 6

X

7 3 9
0 5
8 1
4 2 6

Review Exercises

1. Name and classify this angle.

2. Name and classify this angle.

3. Name and classify this angle.

4. What kind of lines are these?

HELPFUL HINTS

Polygons are closed figures made up of line segments.

triangle	rectangle	square	parallelogram	trapezoid
3 sides	4 sides, 4 right angles	4 congruent sides, 4 right angles	4 sides, opposite sides parallel	4 sides, 1 pair of parallel sides

1	
2	
3	
4	
5	
6	
7	
8	
9	
10	
Score	

Name each polygon. Some may have more than one name.

S.

S.

1.

2.

3.

4.

5.

6.

7.

8.

9.

10.

Problem Solving

A man earned $3.75 per hour. How much was his pay if he worked 4 hours?

73

It is illegal to photocopy this page. Copyright © 2023, Richard W. Fisher

Review Exercises	Speed Drills

1. 3) 75   2. 62
                       x 50

3. 732   4. 723
      46   - 435
  + 322

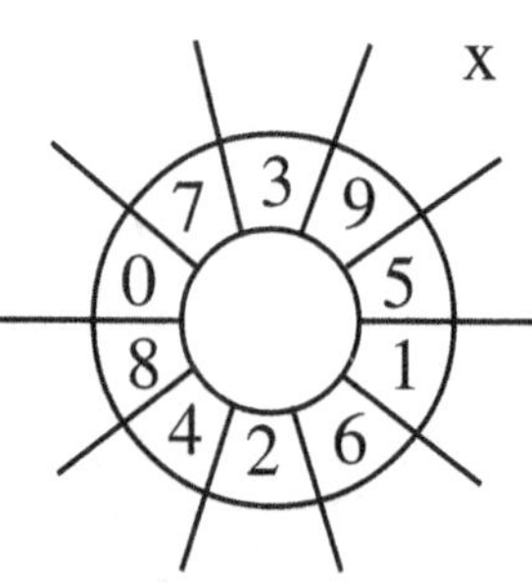

Triangles can be classified by sides and angles.

Sides

equilateral — 3 congruent sides
scalene — no congruent sides
isoceles — 2 congruent sides

Angles

acute — 3 acute angles
right — 1 right angle
obtuse — 1 obtuse angle

HELPFUL HINTS

Classify each triangle by its sides and angles.

S. sides: ______ angles: ______
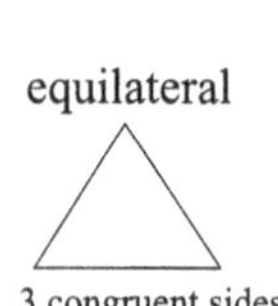

S. sides: ______ angles: ______
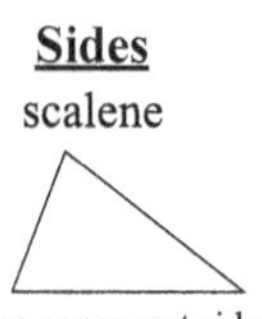

1. sides: ______ angles: ______
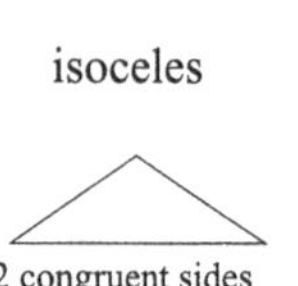

2. sides: ______ angles: ______
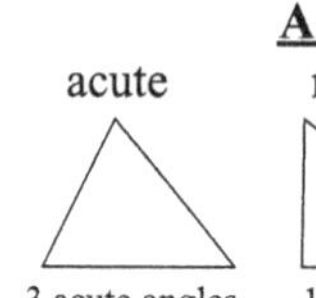

3. sides: ______ angles: ______
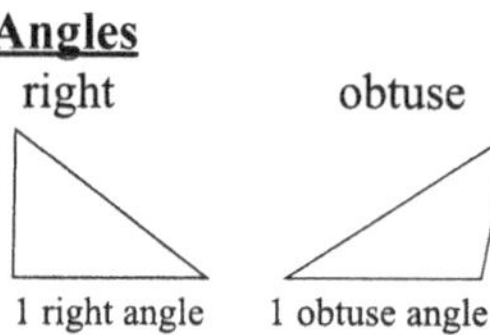

4. sides: ______ angles: ______

5. sides: ______ angles: ______
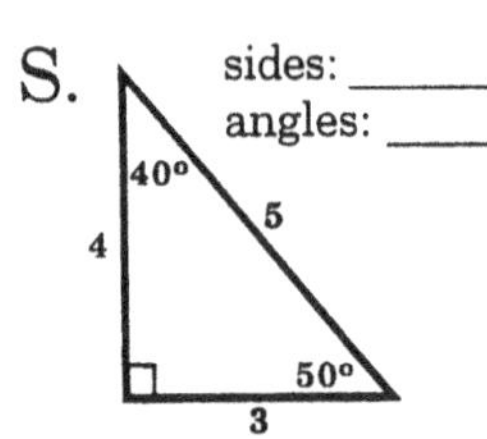

6. sides: ______ angles: ______
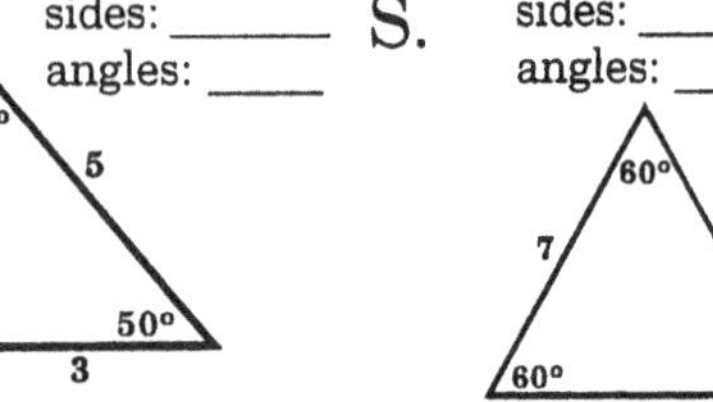

7. sides: ______ angles: ______
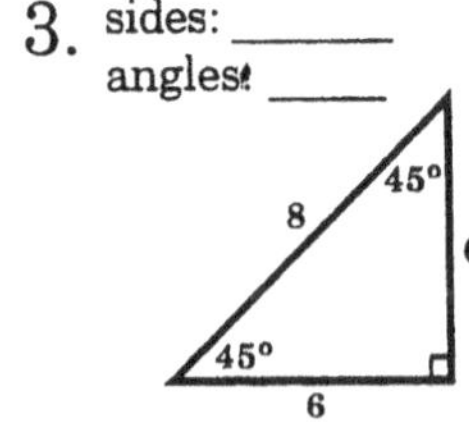

8. sides: ______ angles: ______
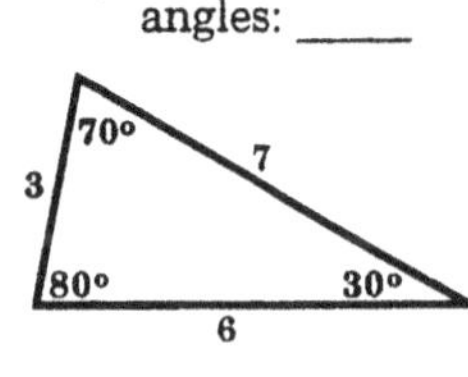

9. sides: ______ angles: ______
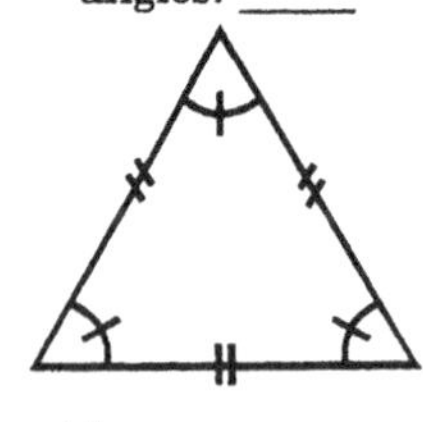

10. sides: ______ angles: ______
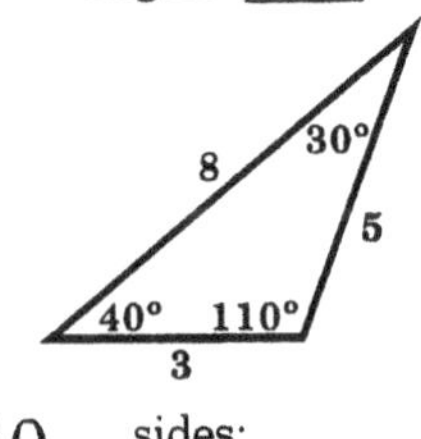

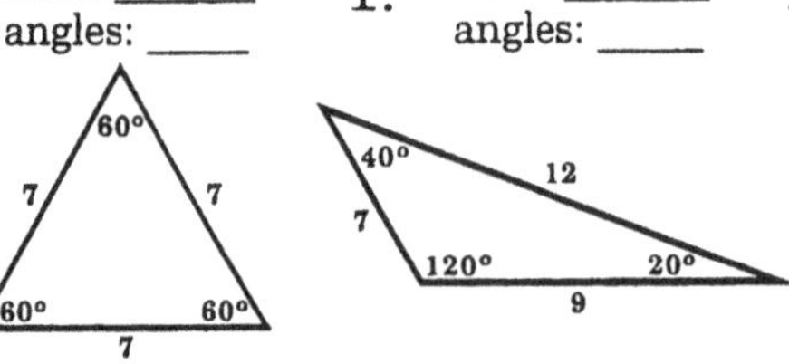
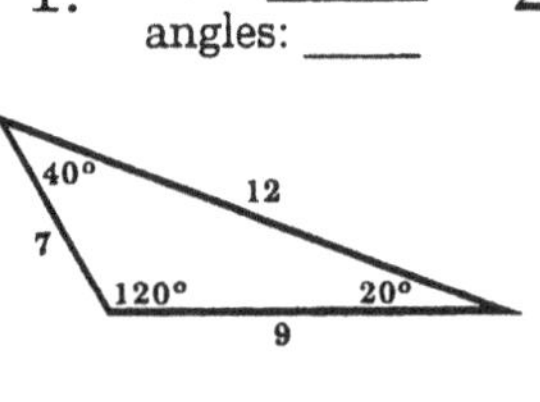
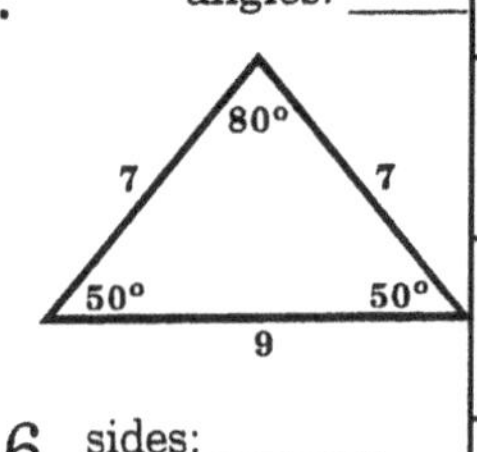
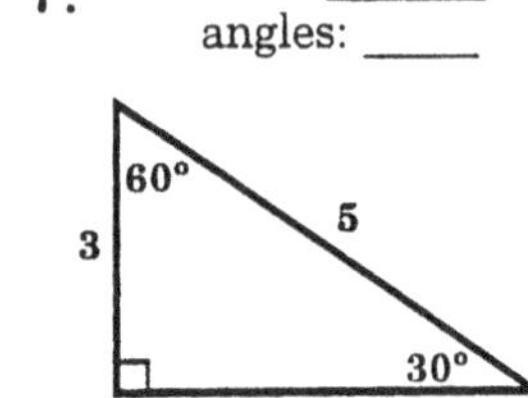
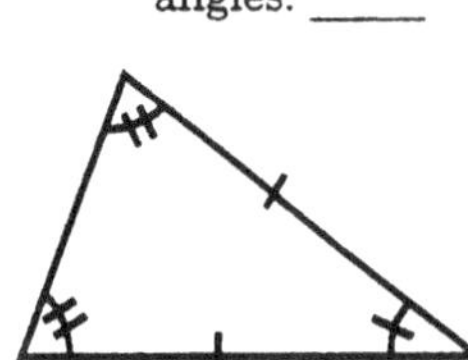
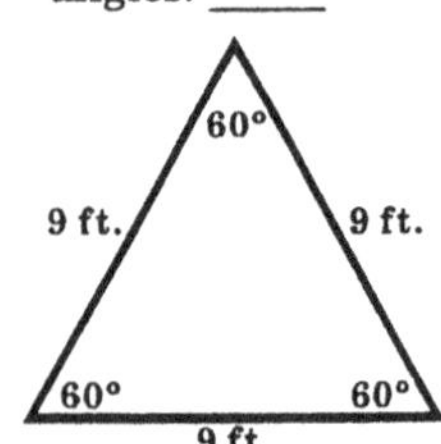
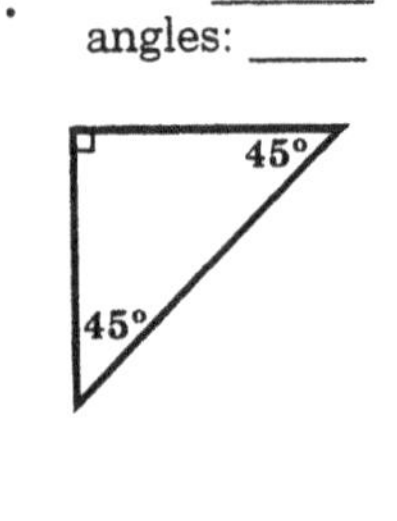

1	
2	
3	
4	
5	
6	
7	
8	
9	
10	

Score

74   Buses hold 60 people. How many buses are needed for 143 people?

Problem Solving

It is illegal to photocopy this page. Copyright © 2023, Richard W. Fisher

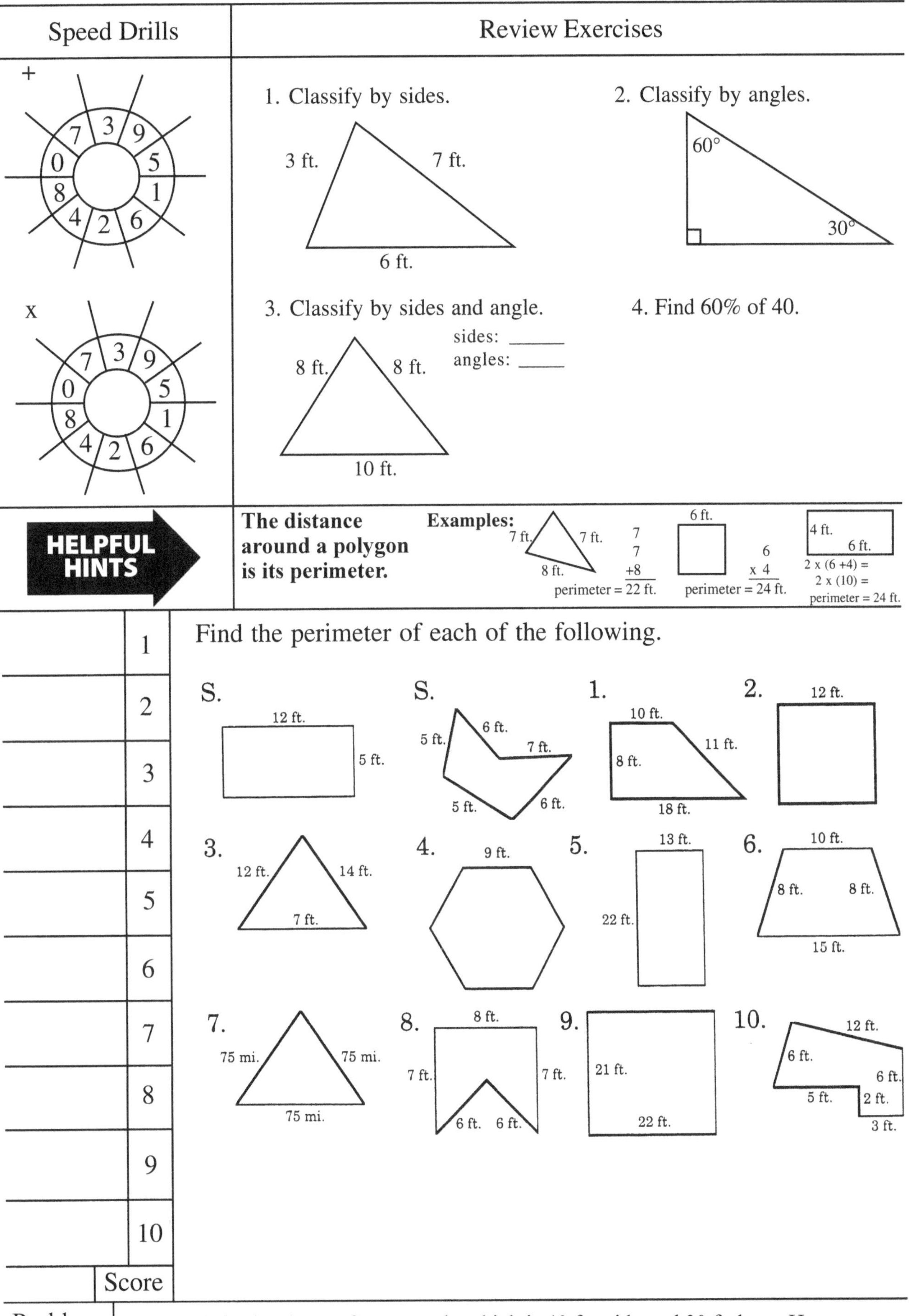

A yard is in the shape of a rectangle which is 40 ft. wide and 30 ft. long. How many feet of fence will it take to go all around the yard? (Hint: Draw a sketch.)

75

It is illegal to photocopy this page. Copyright © 2023, Richard W. Fisher   It is illegal to photocopy this page. Copyright © 2008, Richard W. Fis

Review Exercises	Speed Drills

1.   Find the perimeter.   2.   Find the perimeter.

5 ft. [16 ft.]   14 ft.

3.   6 is what % of 30?   4. $\dfrac{3}{4}$

$-\ \dfrac{1}{3}$

These are the parts of a circle.

chord
diameter
radius
center

*** The length of the diameter is twice that of the radius.**

HELPFUL HINTS

Use the figure to answer the following.

Circle A

G   F
C   E
D
B
A

Circle B

x
Y   V
Z   P
O
R S T

S. What part of the circle is CE?

S. Name a chord in circle B.

1. What part of circle A is DF?

2. What part of circle B is VT?

3. Name 2 radii in circle A

4. Name 2 chords in circle A

5. If the length of CE is 16 ft., what is the length of CD?

6. Name the center of circle B.

7. Name 2 chords in circle B.

8. If PS in Circle B is 8 ft., what is the length of XS?

9. name 2 radii in circle B.

10. Name a diameter in circle B.

1	
2	
3	
4	
5	
6	
7	
8	
9	
10	
Score	

Problem Solving

76   A city is in the shape of a square. If its perimeter is 64 miles, what is the length of each side of the city?

It is illegal to photocopy this page. Copyright © 2023, Richard W. Fisher

Speed Drills	Review Exercises

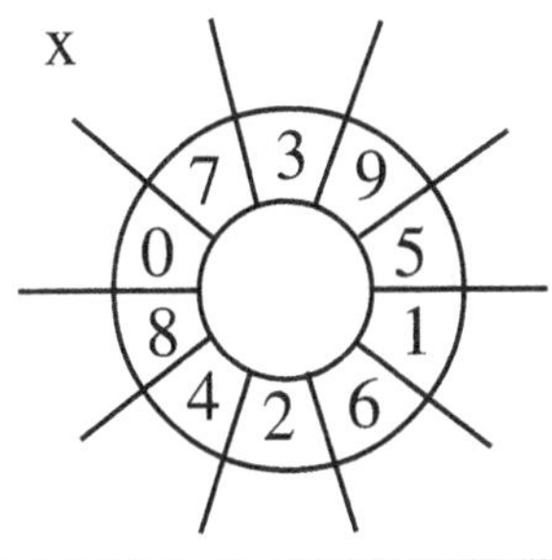

+

1. $233 + 15 + 6 =$ 2.   72
                                              $-\ 16$

3.   Find 20% of 60.   4.   3 is what % of 15?

x

HELPFUL HINTS

The distance around a circle is called its circumference. The Greek letter π = pi = 3.14. To find the circumference, multiply π x diameter. $C = \pi$ x d. Examples:

$C = \pi$ x d
$C = 3.14$ x 6

6 ft.

$\begin{array}{r} 3.14 \\ \times\ \ 6 \\ \hline \end{array}$
18.84 ft

$C = \pi$ x d
$C = 3.14$ x 12

6 ft.

$\begin{array}{r} 3.14 \\ \times\ 12 \\ \hline 6\ 2\ 8 \\ 3\ 1\ 4\ 0 \\ \hline 3\ 7.6\ 8\ \text{ft.} \end{array}$

Find the circumference of each of the following. If there is no figure, draw a sketch.

1	
2	
3	
4	
5	
6	
7	

S.

4 ft.

S.

4 ft.

1.

6 ft.

2.

4 ft.

3. A circle with diameter 5 ft.

4. A circle with radius 1 ft.

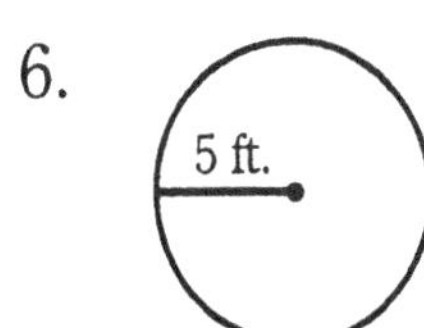

5.

12 ft.

6.

5 ft.

7. A circle with radius 2 ft.

	Score

Problem Solving   A garden is in the shape of a circle. If the diameter is 3 feet, how far is it all the way around the garden?

77

It is illegal to photocopy this page. Copyright © 2023, Richard W. Fisher

Review Exercises	Speed Drills

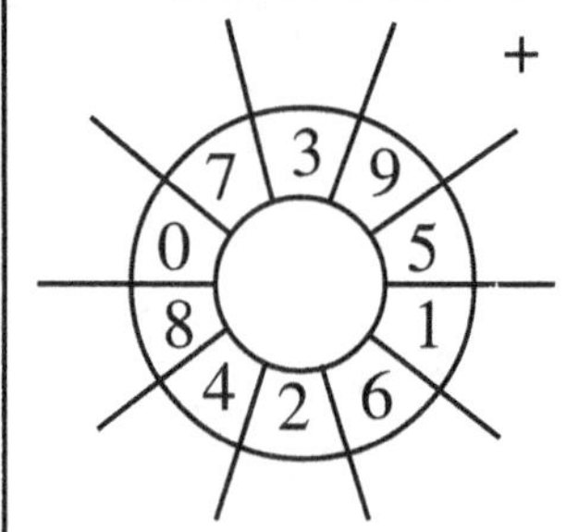

1.   3.2
    + 6.1

2. 12.6 + 23.9 =

3. $2\frac{1}{2}$ x $\frac{1}{2}$ =

4.   3 x 32. 5 =

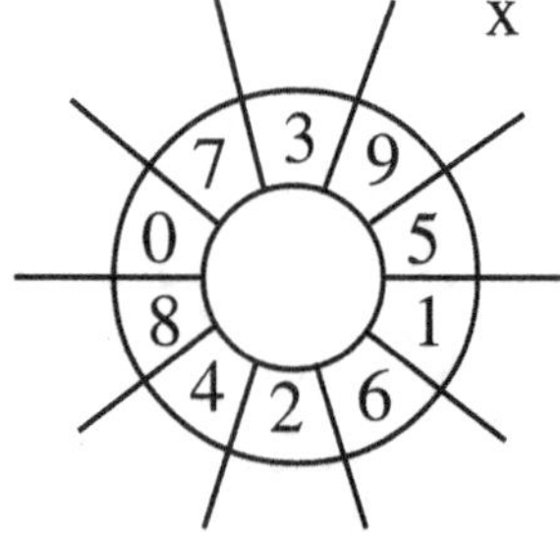

The number of square units needed to cover a region is called its area.

Examples:
area square = side x side　　　　　**area rectangle = length x width**

A= s x s   7
s= 7 ft.　A= 7 x 7   x 7
　　　　　　　49 sq. ft.

w= 7 ft.
A= l x w   12
A= 12 x 7   x 7
l = 12 ft.　　　84 sq. ft.

Hint: 1. Start with formulas   2. Substitute values.   3. Solve the problem.

Find the following areas. If there is no figure, make a sketch.

S. 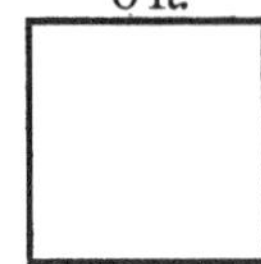
6 ft.

S. 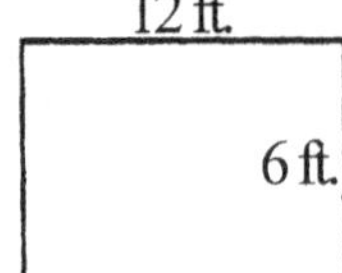
12 ft.
6 ft.

1. 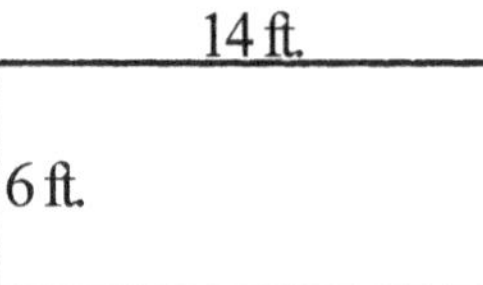
14 ft.
6 ft.

2. 20 ft.

3. A rectangle with length 12 ft. and width 11 ft.

4. 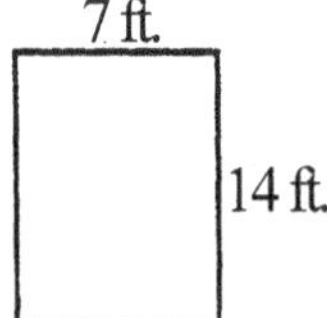
7 ft.
14 ft.

5. 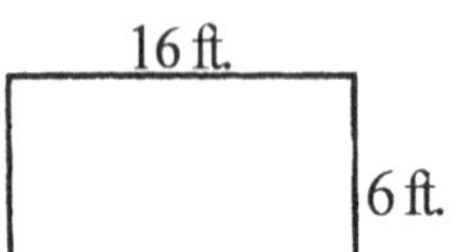
16 ft.
6 ft.

6. 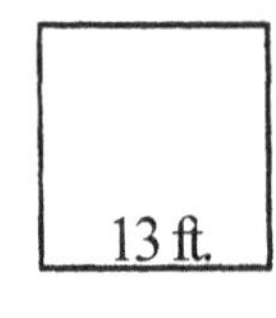
13 ft.

7. A square with sides 11 ft.

1	
2	
3	
4	
5	
6	
7	

Score	

A floor is in the shape of a rectangle. The length is 14 feet and the width is 13 feet. What is the perimeter of the floor?

Problem Solving

It is illegal to photocopy this page. Copyright © 2023, Richard W. Fisher

Review Exercises	Speed Drills

1. Find the perimeter

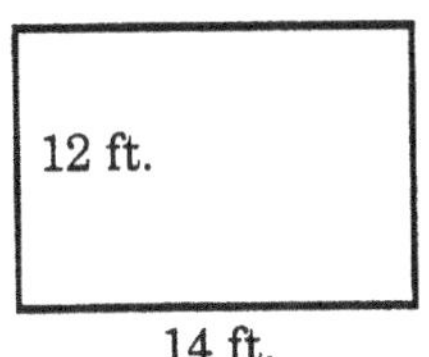

2. Find the area

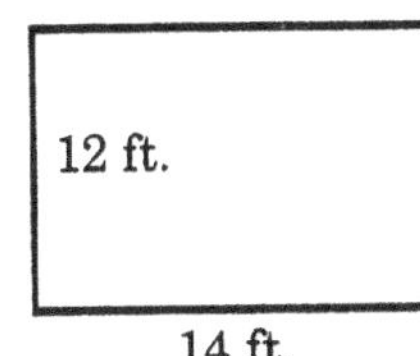

3. Find the circumference

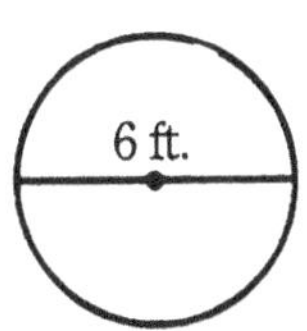

4. Find the area

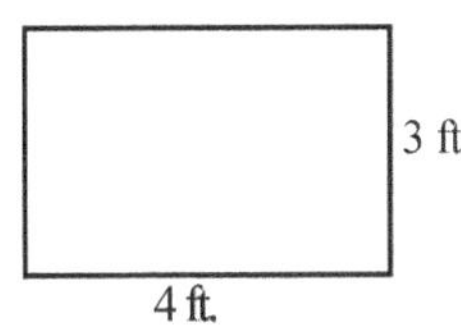

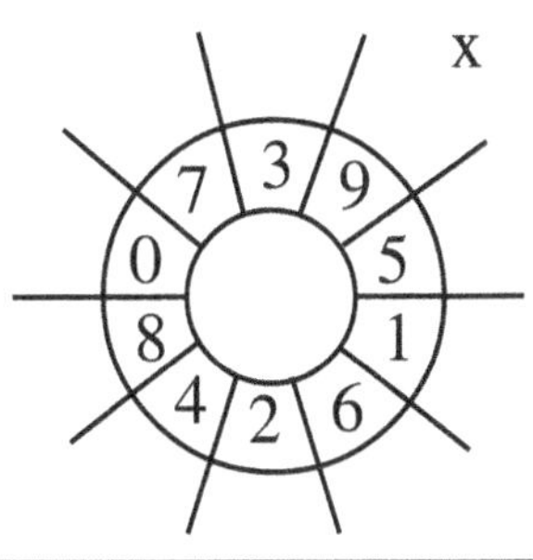

Remember these formulas.   For Areas: 1. Write formula 2. Substitute values 3. Solve problem

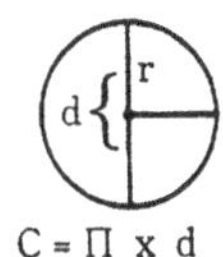
$C = \Pi \times d$

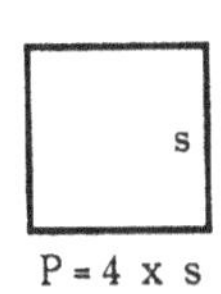
$P = 4 \times s$
$A = s \times s$

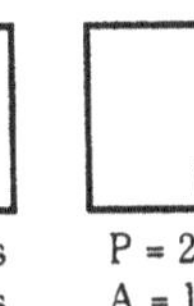
$P = 2 \ (l + w)$
$A = l \times w$

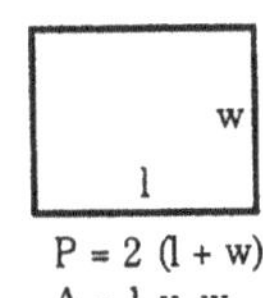
$A = \dfrac{b \times h}{2}$
$P = $ Sum of all sides

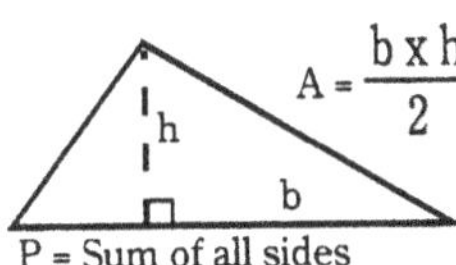

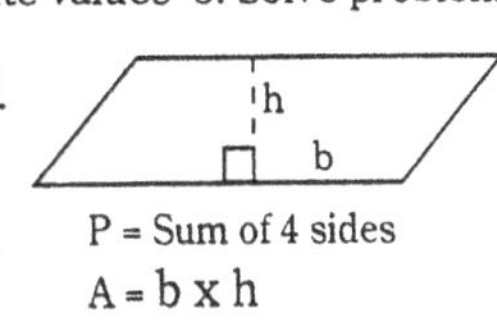
$P = $ Sum of 4 sides
$A = b \times h$

HELPFUL HINTS

]Find the perimeter or circumference. Second, find the area. If there {is no figure, make a sketch.

S.
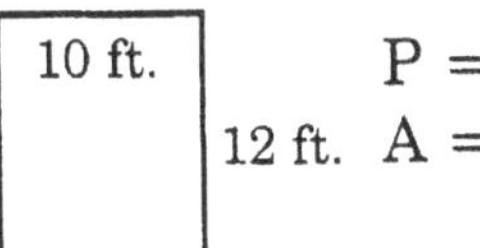

P =   A =

S.

P =
A =

1.

P =
A =

2.
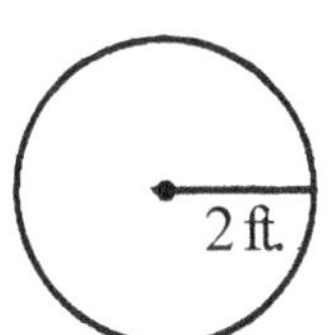

P =
A =

3.
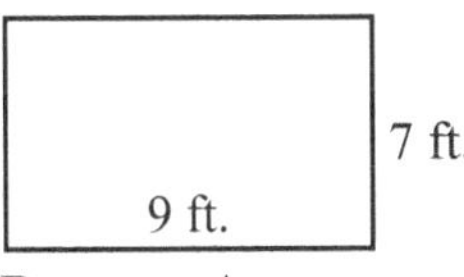

P =   A =

4.
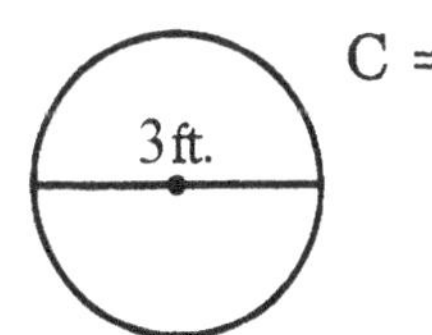

C =

5.

C =

6.
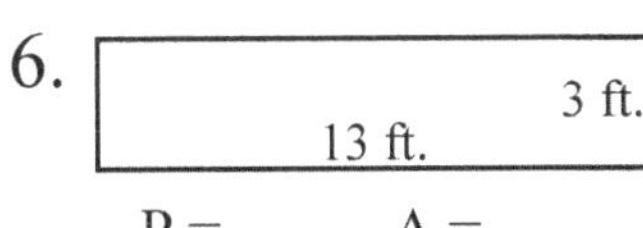

P =   A =

7. A square with sides 8 ft.

P =   A =

1	
2	
3	
4	
5	
6	
7	
Score	

Problem Solving   A man wants to buy a tent which is in the shape of a rectangle. If the length is 18 feet and the width is 12 feet, what is the perimeter of the tent?

It is illegal to photocopy this page. Copyright © 2023, Richard W. Fisher

Speed Drills	Review Exercises

+ (speed drill wheel: 7 3 9 5 1 6 2 4 8 0)

x (speed drill wheel: 7 3 9 5 1 6 2 4 8 0)

HELPFUL HINTS

1.   761
   − 27

2.   33
      22
   + 44

3. $\frac{1}{2}$ of $\frac{2}{3} =$

4. $2\frac{1}{2} \div \frac{1}{2} =$

cube — vertex, edge, face

rectangular prism

triangular prism

triangular pyramid

sphere

square pyramid

cone

cylinder

cones and cylinders do not have straight edges

| 1 |
| 2 |
| 3 |
| 4 |
| 5 |
| 6 |
| 7 |

Identify the shape and the number of each part.

S.   name _______
   faces _______
   edges _______
   vertices _______

S.   name _______
   faces _______
   edges _______
   vertices _______

1.   name _______

2.   name _______
   faces _______
   edges _______
   vertices _______

3.   name _______

4.   name _______
   faces _______
   edges _______
   vertices _______

5.   name _______
   faces _______
   edges _______
   vertices _______

6.   name _______

7.   How many more faces does a cube have than a triangular prism?

Score

80   Jill had $7.35 and spent $5.40. How much money does she have left?

Problem Solving

It is illegal to photocopy this page. Copyright © 2023, Richard W. Fisher

Use the figures to answer 1 - 8

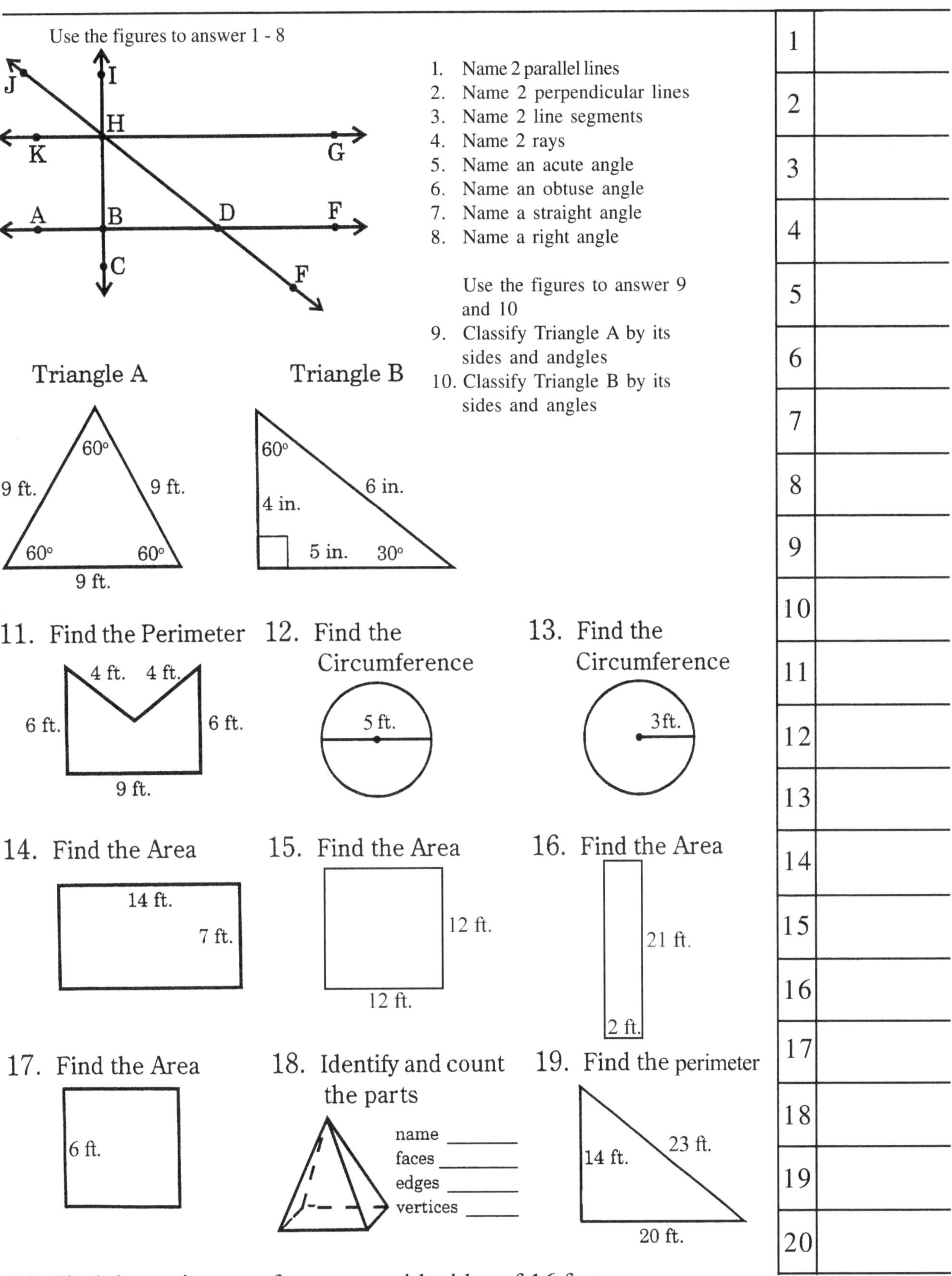

1. Name 2 parallel lines
2. Name 2 perpendicular lines
3. Name 2 line segments
4. Name 2 rays
5. Name an acute angle
6. Name an obtuse angle
7. Name a straight angle
8. Name a right angle

Use the figures to answer 9 and 10

9. Classify Triangle A by its sides and andgles
10. Classify Triangle B by its sides and angles

11. Find the Perimeter

12. Find the Circumference

13. Find the Circumference

14. Find the Area

15. Find the Area

16. Find the Area

17. Find the Area

18. Identify and count the parts

19. Find the perimeter

20. Find the perimeter of a square with sides of 16 feet.

1	
2	
3	
4	
5	
6	
7	
8	
9	
10	
11	
12	
13	
14	
15	
16	
17	
18	
19	
20	

81

It is illegal to photocopy this page. Copyright © 2023, Richard W. Fisher

Speed Drills	Review Exercises

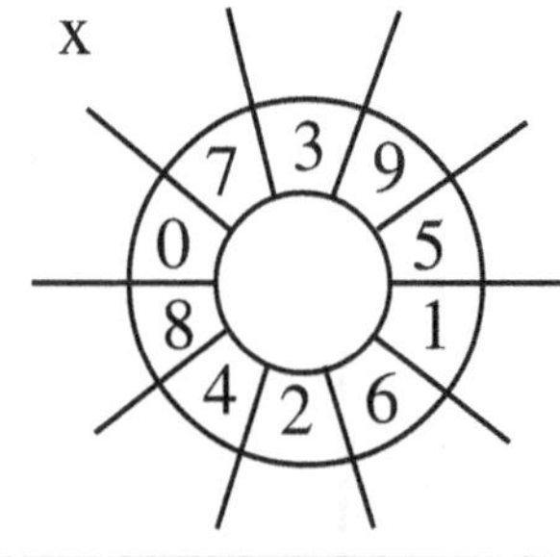

HELPFUL HINTS

1. 703
   − 26

2. $\dfrac{2}{5}$
   $+ \dfrac{1}{2}$

3. 3.6
   + 7.2

4. Find 3% of 80.

The factor of a whole number is a whole number which divides into it evenly, without a remainder.

Example: Find all factors of 20.

1 x 20 = 20   2 x 10 = 20   4 x 5 = 20

So 1, 20, 2, 10, 4, 5 are all factors of 20.

The number 20 can be divided evenly by each of these factors.

1	
2	
3	
4	
5	
6	
7	
8	
9	
10	
Score	

Find all the factors of each number.

S. 10   S. 12   1. 15   2. 16

3. 9   4. 24   5. 30   6. 18

7. 8   8. 25   9. 17   10. 21

How much will a worker earn in 6 hours if he earns $4.50 per hour?

Problem Solving

82

It is illegal to photocopy this page. Copyright © 2023, Richard W. Fisher

| Review Exercises | Speed Drills |

1. 324
    x 6

2. $\frac{3}{4}$
    $-\ \frac{1}{4}$

3. 1.22
    x .3

4. Find the perimeter.
    7 ft.
    3 ft.

The greatest common factor is the largest factor that two numbers have in common.

Example: Find the greatest common factor of 12 and 16

Find the factors   12: 1, 2, 3,④ 6, 12
of each number: 16: 1, 2,④ 8, 16
                      4 is the greatest common factor.

HELPFUL HINTS

Find the greatest common factor of each pair of numbers.

S. 8 and 10   S. 12 and 20   1. 6 and 8   2. 12 and 15

3. 15 and 20   4. 18 and 15   5. 16 and 20   6. 20 and 16

7. 14 and 10   8. 6 and 12   9. 8 and 12   10. 20 and 24

1	
2	
3	
4	
5	
6	
7	
8	
9	
10	
Score	

Problem Solving A boy scored 153 points on a video game. This was 12 points more than his brother. How many points did his brother score?

83

It is illegal to photocopy this page. Copyright © 2023, Richard W. Fisher

Speed Drills	Review Exercises

Speed Drills

+ 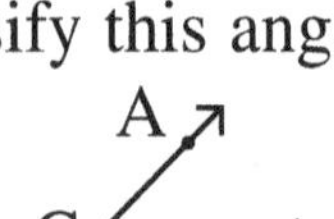

x

HELPFUL HINTS →

Review Exercises

1. Find all factors of 20.

2. Find the area.

12 ft.

5ft.

3. Classify this angle.

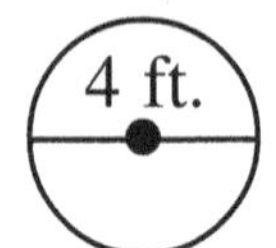

4. Circumference =

4 ft.

A multiple of a number is the product of that number and any whole number.
The multiples of a number can be found by multiplying it by 0, 1, 2, 3, 4, and so on.

Example: Find the first six multiples of 3.
           3:   0,   3,   6,   9,   12,   15
These are found by multiplying 3 by 0, 1, 2, 3, 4, and 5

1	
2	
3	
4	
5	
6	
7	
8	
9	
10	
Score	

Complete the lists of multiples for each number.

S.   2: 0, 2, □,□,□,□   S.   6: □, 6, □,□, 24,□

1.   5: 0, 5, □,□,□,□   2.   3: □, 3, □, 9, □,□

3. 10: □, 10, 20, □,□,□   4.   4: □,□,□, 12, 16, 20

5. 11: 0, 11, □, 33, □, 55   6.   8: 0, 8, 16, □,□,□

7. 20: 0, 20, 40, □,□,□   8.   7: 0, 7, □, 21, □,□

9. 30: 0, 30, 60, □,□,□   10. 9: 0, 9, 18, □, 36, □

84

The Smith family planned a 20 mile hike. The first day they hiked 7 miles, and the second day 5 miles. How many miles are left to hike? | **Problem Solving**

It is illegal to photocopy this page. Copyright © 2023, Richard W. Fisher

Review Exercises	Speed Drills

1. List all factors of 12.   2. 2 is what % of 5?

3. Find the area: 4. List the first five multiples of 3.

7 ft.

15 ft.

The least common multiples of two whole numbers is the smallest whole number, other than 0, that they both divide into evenly.

Examples: The least common multiple of:
2 and 3 is 6;   4 and 6 is 12;   3 and 9 is 9

HELPFUL HINTS

Find the least common multiple of each pair of numbers.

S. 3 and 4   S. 6 and 8   1. 3 and 5

2. 4 and 10   3. 5 and 6   4. 4 and 12

5. 2 and 6   6. 6 and 9   7. 5 and 10

8. 7 and 5   9. 2 and 8   10. 15 and 10

1	
2	
3	
4	
5	
6	
7	
8	
9	
10	
Score	

Problem Solving | A girl bought a bag of chips for $.34, a soda for $.25, and a candy bar for $.20. How much did she spend altogether?

85

It is illegal to photocopy this page. Copyright © 2023, Richard W. Fisher

Speed Drills　　　　　Review Exercises

+

1.　3.6
　　5.0
　+ 3.2

2.　3.72
　- 1.45

3.　2.4
　x　3

4.　$2\overline{)\,.46}$

x

HELPFUL HINTS

A number sentence that has an equal sign, =, is called an equation. An equation may contain a variable, which is a letter used to represent an unspecified number. When you replace the variable with a number which makes the equation true, you have solved the equation.

Examples: Solve the following equations:

$$x + 2 = 7 \qquad 15 + x = 23 \qquad y + 10 = 20$$
$$5 + 2 = 7 \qquad 15 + 7 = 23 \qquad 10 + 20 = 20$$
$$\text{so } x = 5 \qquad \text{so } x = 7 \qquad \text{so } y = 10$$

1	
2	
3	
4	
5	
6	
7	
8	
9	
10	
Score	

Solve each of the following equations:

S.　$x + 7 = 13$　　　　S.　$x = 5 + 12$　　　　1.　$8 + x = 12$

2.　$x + 4 = 10$　　　　3.　$13 = 5 + y$　　　　4.　$12 = y + 4$

5.　$x + 12 = 23$　　　　6.　$20 = 13 + x$　　　　7.　$13 = m + 3$

8.　$4 + m = 12$　　　　9.　$16 = 2 + m$　　　　10.　$n + 12 = 30$

A playing field is in the shape of a square. If each side is 80 feet long, how many feet is it around the playing field?

Problem Solving

86

It is illegal to photocopy this page. Copyright © 2023, Richard W. Fisher

Review Exercises	Speed Drills

1. Find the least common multiple of 4 and 6.

2. Find the greatest common factor of 20 and 15.

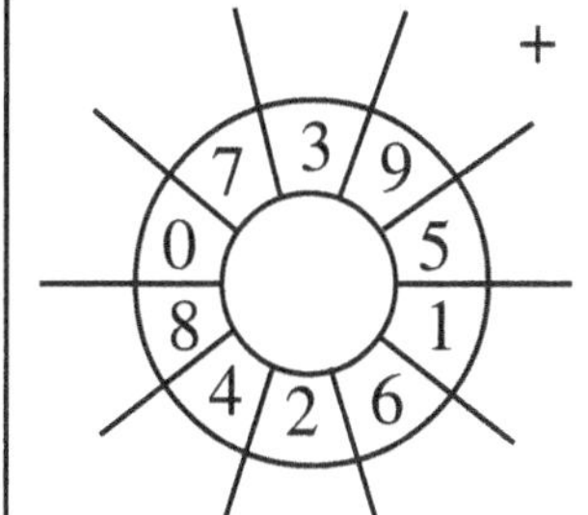

3. $2\overline{)\,.33}$

4.
$$\begin{array}{r} .3 \\ .4 \\ +\ .4 \\ \hline \end{array}$$

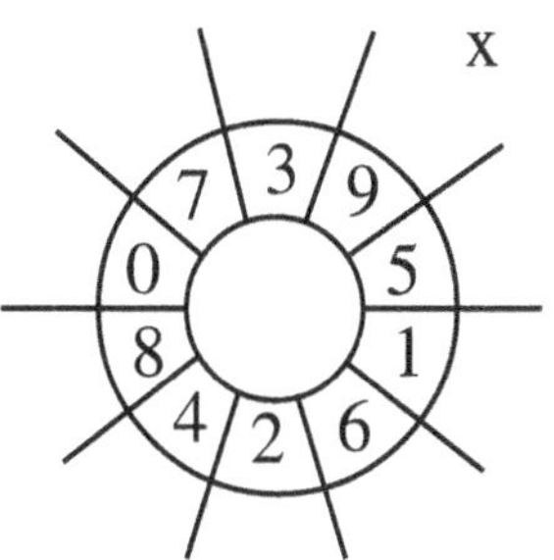

Remember, when you replace the variable in an equation with the number which makes the equation true, you have solved the equation.

Equations may contain addition, subtraction, multiplication or division.

HELPFUL HINTS

Solve each of the following equations:

S. $x - 2 = 5$ S. $6 = 12 - n$ 1. $n - 2 = 5$

2. $n - 5 = 4$ 3. $6 - x = 2$ 4. $2 = 5 - n$

5. $10 = n - 4$ 6. $8 - 3 = n$ 7. $9 - x = 3$

8. $20 - n = 15$ 9. $n - 7 = 18$ 10. $x - 3 = 11$

	Score
1	
2	
3	
4	
5	
6	
7	
8	
9	
10	

Problem Solving | A spelling test had 20 words. If Tran got 80% of the words correct, how many words did she get correct?

87

It is illegal to photocopy this page. Copyright © 2023, Richard W. Fisher

Speed Drills	Review Exercises

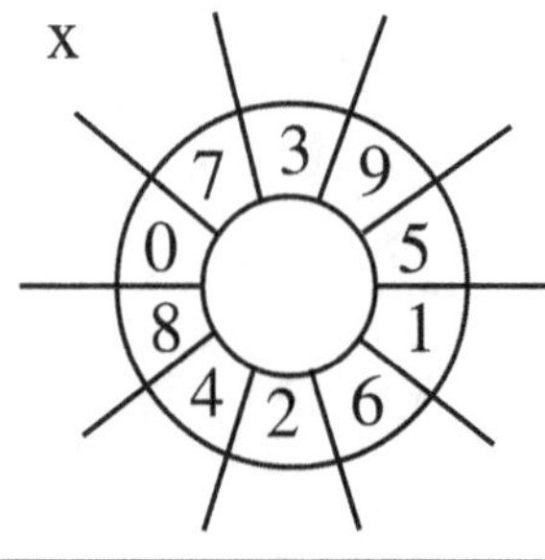

1. $\dfrac{2}{5} \div \dfrac{1}{4}$

2. $\dfrac{2}{5} \times \dfrac{1}{3}$

3. $\begin{array}{r} \dfrac{5}{8} \\[4pt] -\dfrac{1}{8} \\ \hline \end{array}$

4. $\begin{array}{r} \dfrac{3}{8} \\[4pt] -\dfrac{2}{8} \\ \hline \end{array}$

When a number is written right beside a variable, this means to <u>multiply</u>. The variable should be on the left.

$$3n = 15 \qquad\qquad 5n = 20$$
$$\text{means} \qquad\qquad \text{means}$$
$$3 \times n = 15 \qquad\qquad 5 \times n = 20$$
$$\boxed{n = 5} \qquad\qquad \boxed{n = 4}$$

HELPFUL HINTS ➤

	1
	2
	3
	4
	5
	6
	7
	8
	9
	10
Score	

Solve each of the following equations:

S. $2n = 12$ S. $3n = 24$ 1. $5n = 25$

2. $2n = 24$ 3. $12 = 3n$ 4. $28 = 4n$

5. $4n = 32$ 6. $5n = 35$ 7. $36 = 4n$

8. $7n = 35$ 9. $20 = 2n$ 10. $3n = 33$

A girl bought $3.24 worth of groceries. If she paid with a $5.00 bill, how much change should she receive? | **Problem Solving**

It is illegal to photocopy this page. Copyright © 2023, Richard W. Fisher

Review Exercises	Speed Drills

1. Classify this angle.

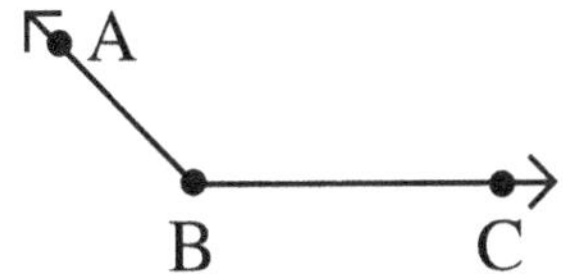

2. $6\,\overline{)127}$

3. $\begin{array}{r} 35 \\ \times\ 23 \\ \hline \end{array}$

4. $234 + 14 + 123 =$

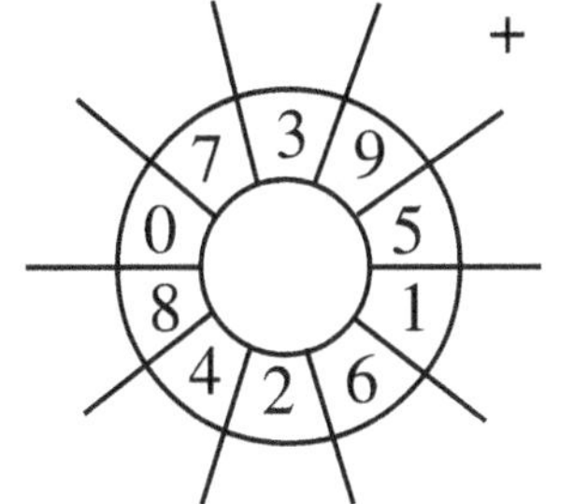

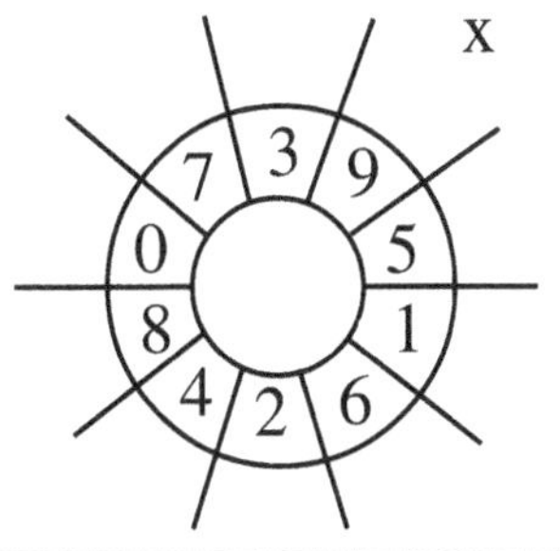

Equations with division may be written in two ways.

Examples:

$$x \div 2 = 5$$
$$10 \div 2 = 5$$
$$\boxed{x = 10}$$

$$\frac{n}{2} = 3$$

This means $n \div 2 = 3$
$$6 \div 2 = 3$$
$$\boxed{n = 6}$$

Solve each of the following equations:

S. $\dfrac{x}{3} = 4$

S. $x \div 2 = 4$

1. $x \div 3 = 5$

2. $n \div 4 = 3$

3. $\dfrac{n}{2} = 4$

4. $\dfrac{12}{n} = 4$

5. $\dfrac{12}{2} = n$

6. $\dfrac{n}{7} = 3$

7. $\dfrac{n}{2} = 10$

8. $\dfrac{15}{x} = 3$

9. $\dfrac{n}{3} = 7$

10. $\dfrac{n}{4} = 6$

1	
2	
3	
4	
5	
6	
7	
8	
9	
10	
Score	

Problem Solving   An hour is 60 minutes. If a television program is $1\frac{1}{2}$ hours long, how many minutes long is the program?

89

It is illegal to photocopy this page. Copyright © 2023, Richard W. Fisher

Speed Drills	Review Exercises

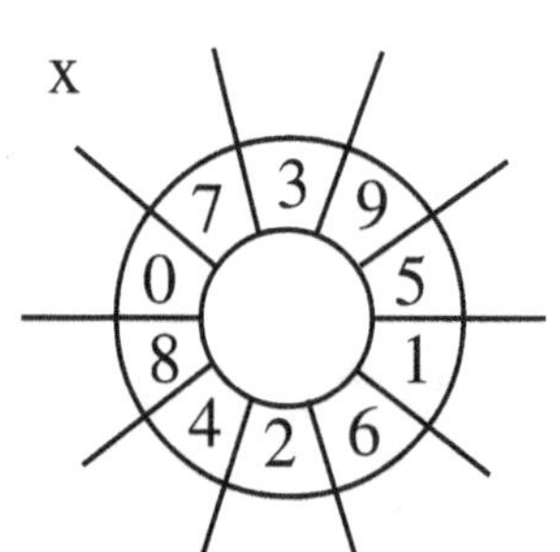

1. Find the perimeter.

5 ft.

3 ft. [ ]

2. Find the area.

6 ft.

4 ft. [ ]

3. Find the circumference.

3 ft.

4. List all the factors of 20.

HELPFUL HINTS ➤ **Use what you have learned to solve each of the following equations.**

1	
2	
3	
4	
5	
6	
7	
8	
9	
10	
Score	

Solve each of the following equations:

S. $\dfrac{x}{3} = 2$ S. $12 - n = 3$ 1. $x + 2 = 7$

2. $n - 7 = 2$ 3. $4n = 12$ 4. $n \div 2 = 5$

5. $13 = n + 10$ 6. $5 - n = 3$ 7. $18 = 3n$

8. $\dfrac{n}{2} = 7$ 9. $12 \div n = 6$ 10. $5 = 6 - n$

If hamburgers are \$.72 each, how much would 5 of them cost? | **Problem Solving**

It is illegal to photocopy this page. Copyright © 2023, Richard W. Fisher

Find all factors of each number.

1. 20   2. 24   3. 30

Find the greatest common factor of each pair of numbers.

4. 16 and 10   5. 8 and 12   6. 12 and 15

Complete the list of multiples of each number.

7. 2: 0, 2, 4, ☐, ☐, ☐

8. 5: 0, ☐, 10, ☐,☐,☐

9. 7: ☐,☐,☐,☐,☐ , 35

Find the least common multiple of each of the pair of numbers.

10. 3 and 4   11. 4 and 6   12. 3 and 9

Solve each of the following equations.

13. $x + 2 = 5$ 14. $n - 2 = 5$ 15. $3n = 12$

16. $\dfrac{x}{2} = 4$ 17. $6 = n + 4$ 18. $7 - n = 3$

19. $n \div 5 = 2$ 20. $2n = 8$

91

It is illegal to photocopy this page. Copyright © 2023, Richard W. Fisher

Speed Drills

+

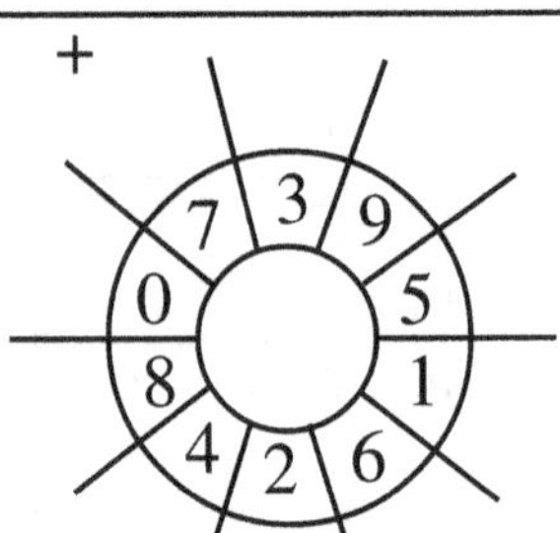

x

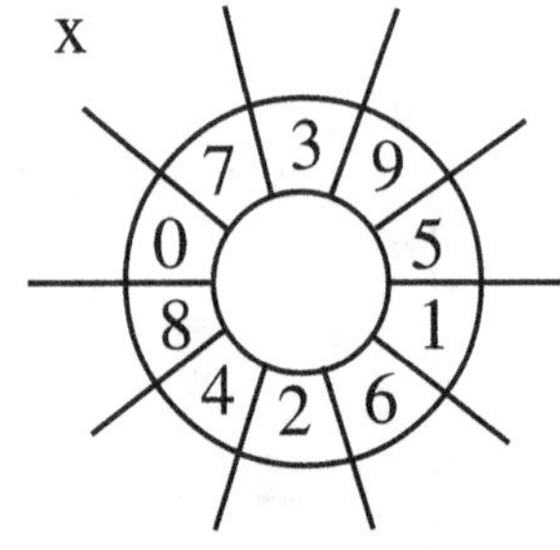

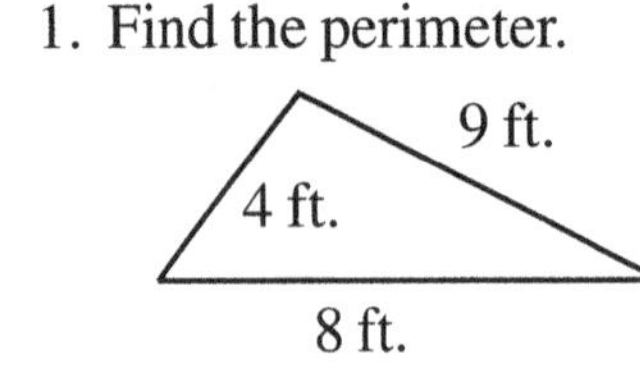

Review Exercises

1. Find the perimeter.

9 ft.
4 ft.
8 ft.

2. $\dfrac{1}{2} \times 1\dfrac{1}{3} =$

3. $\begin{array}{r} \dfrac{1}{4} \\[4pt] +\ \dfrac{1}{3} \end{array}$

4. $\begin{array}{r} \dfrac{4}{5} \\[4pt] -\ \dfrac{1}{2} \end{array}$

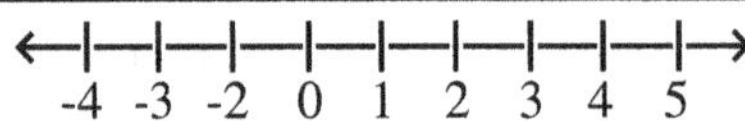

-4 -3 -2 0 1 2 3 4 5

Integers to the left of zero are negative and less than zero. Integers to the right of zero are positive and greater than zero. When two integers are on a number line, the one farthest to the right is greater. Think of positive integers as money you have. Think of negative integers as money you owe.

Example:

-3 + -4 = -7	-3 + 5 = 2	3 + -7 = -4
You owe 3 and	You have 5.	You have 3.
you owe 4. So,	You owe 3.	You owe 7.
you owe 7.	So, you have 2	So, you owe 4.

HELPFUL HINTS

	1
	2
	3
	4
	5
	6
	7
	8
	9
	10
	Score

Solve each of the following:

S. -2 + -3 =   S. -6 + 2 =   1. 5 + -4 =

2. -5 + 4 =   3. -5 + -4 =   4. 9 + -2 =

5. -7 + 2 =   6. -6 + -5 =   7. -7 + -4 =

8. -3 + 9 =   9. 4 + -6 =   10. 12 + -15 =

92

600 people attended a concert. 30% of the people had reserved seats. How many had reserved seats?

Problem Solving

It is illegal to photocopy this page. Copyright © 2023, Richard W. Fisher

Review Exercises	Speed Drills

1. Write 2 to 5 in two other forms.

2. $\dfrac{6}{n} = \dfrac{12}{6}$

3. $3\overline{)\,.132}$

4. $\begin{array}{r} 6.2 \\ \times\ \ 3 \\ \hline \end{array}$

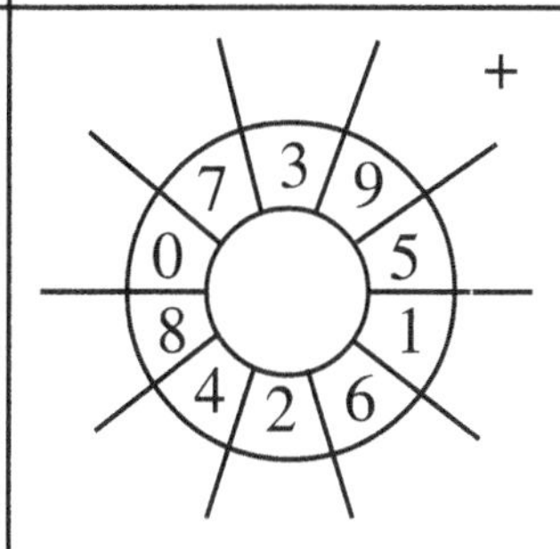

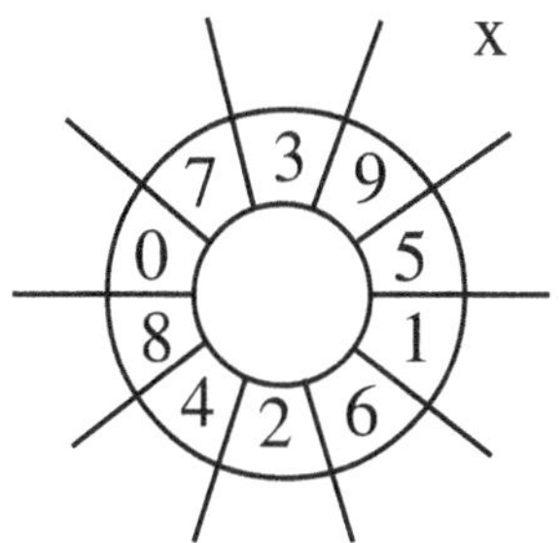

Use what you have learned to solve the following problems.

Think of positive integers as money you <u>have</u>.

Think of negative integers as money you <u>owe</u>.

HELPFUL HINTS

Solve each of the following.

S. -8 + 7 =	S. 8 + -7 =	1. -8 + -7 =	
2. -15 + 13 =	3. 16 + -8 =	4. -7 + -5 =	
5. -6 + -10 =	6. -7 + 12 =	7. 15 + -5 =	
8. -6 + -8 =	9. -20 + 18 =	10. -25 + 27 =	

1	
2	
3	
4	
5	
6	
7	
8	
9	
10	
Score	

Problem Solving | A team played 15 games and lost 3 of them. What percent did they lose?

93

It is illegal to photocopy this page. Copyright © 2023, Richard W. Fisher

Speed Drills	Review Exercises

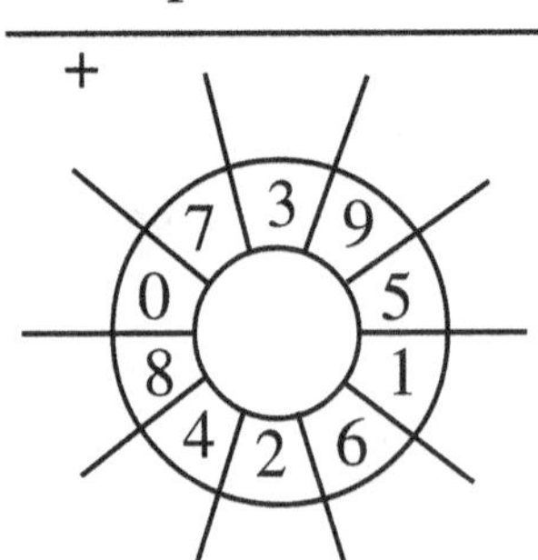

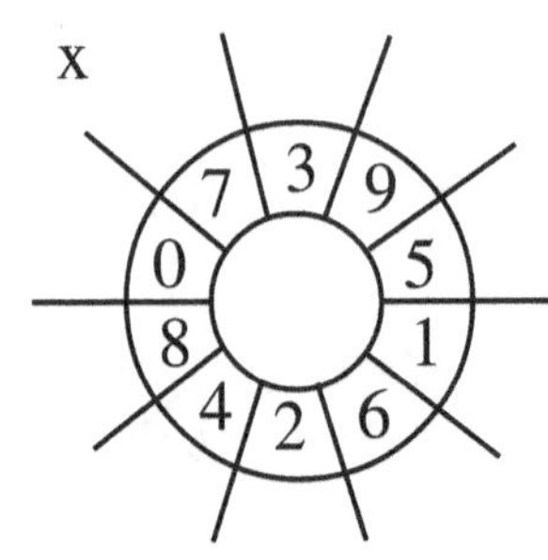

1. Find the perimeter of a square with sides 8 feet.

2. $-6 + 8 =$

3. $-7 + 5 =$

4. Solve the equation.
   $n + 8 = 10$

To subtract integers means to add the opposite.
Examples:

* **a double negative sign - - is the same as +**
- -7 = 7

2 - 5	-2 - 3	6 - -7
means	means	means
2 + -5 = (-3)	-2 + -3 = (-5)	6 + 7 = (13)

HELPFUL HINTS

1	
2	
3	
4	
5	
6	
7	
8	
9	
10	
Score	

Solve each of the following.

S. $-5 - 2 =$ S. $3 - -2 =$ 1. $2 - 7 =$

2. $7 - 8 =$ 3. $-3 - 5 =$ 4. $3 - -2 =$

5. $4 - 6 =$ 6. $-3 - 4 =$ 7. $-3 - -4 =$

8. $-8 - 5 =$ 9. $3 - 8 =$ 10. $2 - -5 =$

94

A business needs 120 postcards to mail to customers. If postcards come in boxes of 20, how many boxes does the business need?

Problem Solving

It is illegal to photocopy this page. Copyright © 2023, Richard W. Fisher

Review Exercises	Speed Drills

1.   -2 - 3 =   2.   2 - 5 =

3.   -3 - -4 =   4.   2 + -7 =

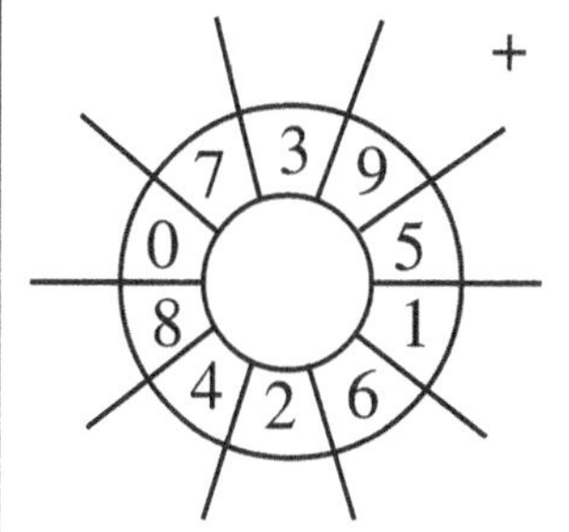

Use what you have learned to solve the following problems.

Think of positive integers as money you <u>have</u>.
Think of negative integers as money you <u>owe</u>.

Subtraction means to add the opposite.   **- - means +**
Examples:
      2 - 3 means
      2 + -3   7 - -2 = 7 + 2

HELPFUL HINTS

Solve each of the following.

S. -8 + 6 =   S. 3 - 7 =   1. -7 + -2 =

2. 6 + -4 =   3. -7 - 3 =   4. 3 - -6 =

5. 6 + -10 =   6. -6 + -10 =   7. -6 + 10 =

8. 3 - 4 =   9. -3 - 4 =   10. -3 - -4 =

1	
2	
3	
4	
5	
6	
7	
8	
9	
10	
Score	

Problem Solving	If eggs cost $1.13 per dozen, how much will 4 dozen cost?

95

It is illegal to photocopy this page. Copyright © 2023, Richard W. Fisher

Speed Drills	Review Exercises

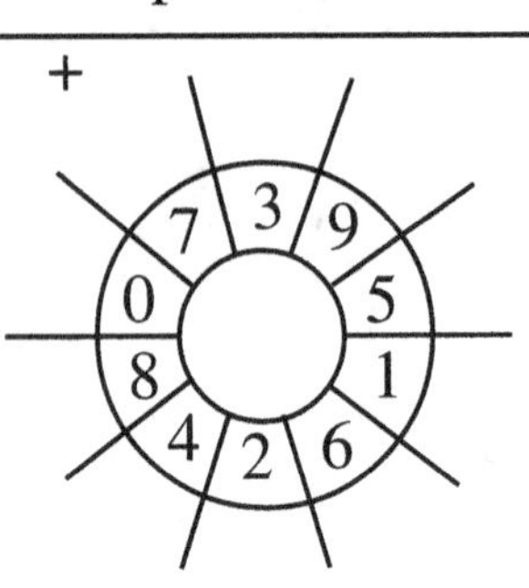

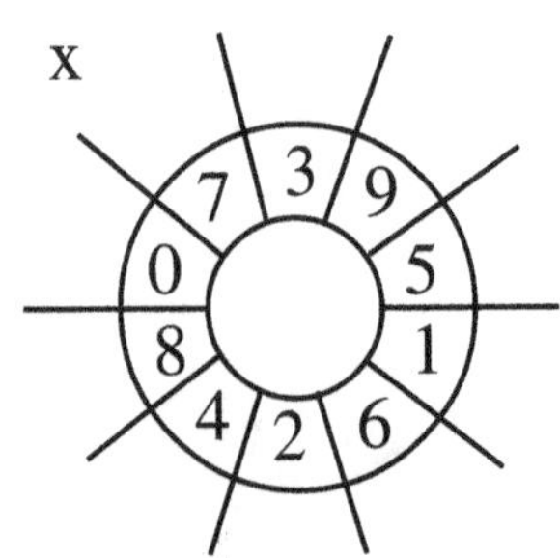

Review Exercises

1. $2 \times \dfrac{3}{4} =$

2. $2\dfrac{1}{2} \div \dfrac{1}{2} =$

3. $\dfrac{1}{3} \times 4 =$

4. $\dfrac{2}{5} \div \dfrac{1}{10} =$

The product of two integers with the same sign is positive.

Example: $-4 \times -3 = 12$

The product of two integers with different signs is negative.

Example: $3 \times -2 = -6$

HELPFUL HINTS

1	
2	
3	
4	
5	
6	
7	
8	
9	
10	
Score	

Solve each of the following.

S. $-2 \times -6 =$ S. $-5 \times 3 =$ 1. $3 \times -7 =$

2. $-5 \times 7 =$ 3. $-5 \times -8 =$ 4. $-4 \times -11 =$

5. $-6 \times 6 =$ 6. $7 \times -4 =$ 7. $-3 \times -8 =$

8. $-13 \times -3 =$ 9. $-11 \times 7 =$ 10. $102 \times -3 =$

96

If the temperature at midnight was 22° and by 6 A.M. the temperature dropped another 15°. What was the temperature at 6 A.M.?

Problem Solving

It is illegal to photocopy this page. Copyright © 2023, Richard W. Fisher

Review Exercises	Speed Drills

1. $5 \overline{)\,106}$ 2. $-3 \times -4 =$

3. $7 + -6 =$ 4. $-3 - 7 =$

The quotient of two integers with the same sign is positive.

Example: $\dfrac{-12}{-2} = 6$

The quotient of two integers with different signs is negative.

Example: $12 \div -3 = -4$

HELPFUL HINTS

Solve each of the following:

S. $-8 \div 2 =$ S. $\dfrac{-10}{-5} =$ 1. $-6 \div -3 =$

2. $-15 \div 3 =$ 3. $18 \div -2 =$ 4. $\dfrac{-20}{-4} =$

5. $\dfrac{24}{-3} =$ 6. $-36 \div -4 =$ 7. $16 \div -2 =$

8. $-15 \div -5 =$ 9. $35 \div -5 =$ 10. $-18 \div 6 =$

1	
2	
3	
4	
5	
6	
7	
8	
9	
10	
Score	

Problem Solving A girl earned 75 dollars a week for 5 weeks. How much money did she earn altogether?

97

It is illegal to photocopy this page. Copyright © 2023, Richard W. Fisher

Reviewing All Integer Operations

1. $-7 + 6 =$

2. $7 + -6 =$

3. $-7 + -6 =$

4. $-9 + -6 =$

5. $8 + -2 =$

6. $-7 + -2 =$

7. $2 - 6 =$

8. $2 - -7 =$

9. $-3 - 5 =$

10. $-5 - -4 =$

11. $8 - 3 =$

12. $3 - 4 =$

13. $2 \times -6 =$

14. $-3 \times -5 =$

15. $-5 \times 7 =$

16. $8 \div -2 =$

17. $-14 \div -2 =$

18. $-24 \div 3 =$

19. $\dfrac{24}{-2} =$

20. $\dfrac{-15}{-5} =$

It is illegal to photocopy this page. Copyright © 2023, Richard W. Fisher

Review Exercises	Speed Drills

1. Find the circumference.   2. Find 12% of 30.

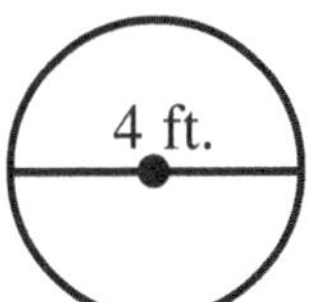

3. $\dfrac{2}{3}$ x $1\dfrac{1}{2}$ =   4. $\dfrac{3}{4} \div \dfrac{1}{4}$ =

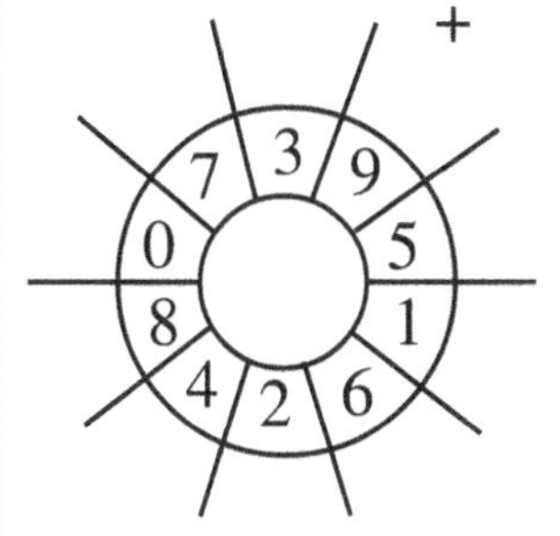

Bar graphs are used to compare information.

1. Read the title.
2. Understand the meaning of the numbers. Estimate if necessary.
3. Study the data.
4. Answer the questions, showing work if necessary.

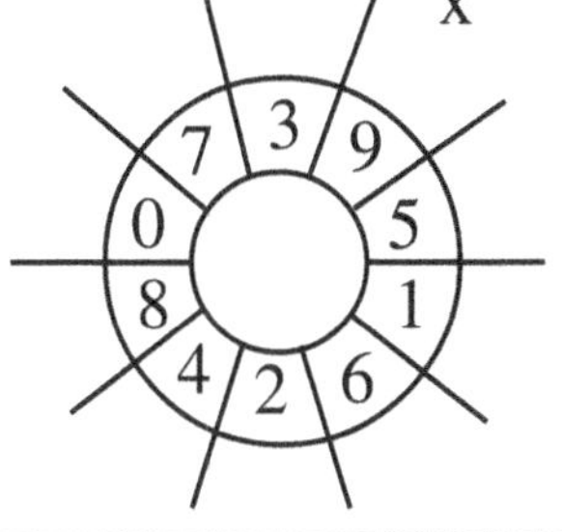

HELPFUL HINTS

Use the information in the graph to answer the questions.

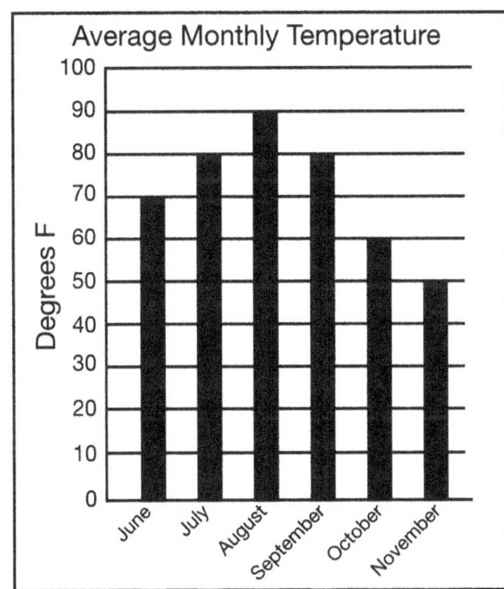

S. Which month had the highest average temperature?

S. What was the average temperature in September?

1. In which month was the average temperature 50°?

2. Which months had the second highest average temperature?

3. Which month had the lowest average temperature?

4. What was the average temperature for July?

5. Which month had the second lowest temperature?

6. How many degrees higher was the average temperature in July than June?

7. How many degrees cooler was the average temperature in October than in September?

8. How many months had higher temperatures than June?

9. How many months had average temperatures lower than July?

10. Which two months had the same average temperature?

1	
2	
3	
4	
5	
6	
7	
8	
9	
10	
Score	

Problem Solving   Bill worked 5 days and earned a total of 30 dollars. If he earned the same amount each day, how much did he earn each day?

99

It is illegal to photocopy this page. Copyright © 2023, Richard W. Fisher

Speed Drills	Review Exercises

+

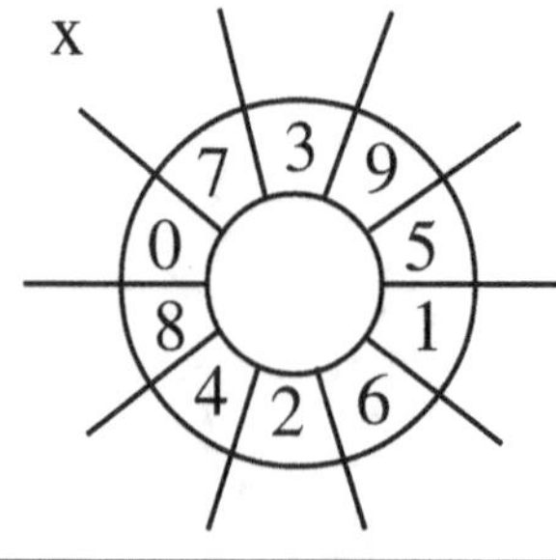

1. Change $\frac{2}{5}$ to a decimal.   2. Change $\frac{3}{5}$ to a percent.

3. Change .3 to a percent.   4. Change .04 to a percent.

x

1. Read the title.
2. Understand the meaning of the numbers.
   Estimate if necessary.
3. Study the data.
4. Answer the questions, showing work if necessary.

HELPFUL HINTS ➤

1	
2	
3	
4	
5	
6	
7	
8	
9	
10	
Score	

100

Use the information in the graph to answer the questions.

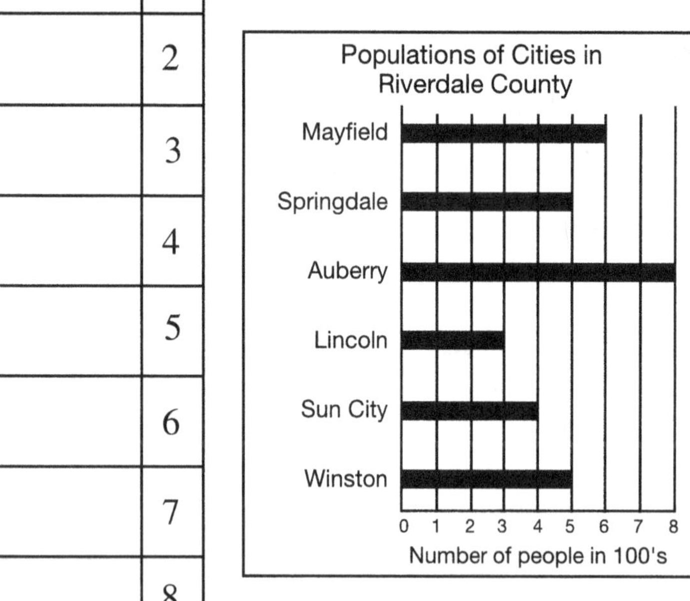

S. Which two cities had the same population?
S. What is the population of Springdale?
1. Which city has the highest population?
2. What is the population of Sun City?
3. What is the total population of Sun City and Winston?
4. How many more people live in Springdale than Lincoln?
5. Which city has the second lowest population?
6. Which city has twice as many people as Lincoln?
7. How many cities have a population greater than 500?
8. What is the total population of the two largest cities?
9. What is the population of Winston?
10. How many people must move to Lincoln to increase the population to 500?

In an election Julie received 665 votes. John received 530. How many more votes did Julie receive than John?

Problem Solving

It is illegal to photocopy this page. Copyright © 2023, Richard W. Fisher

Review Exercises

1. Change $\frac{1}{4}$ to a percent.

2. Change .7 to a percent.

3. Find the area.

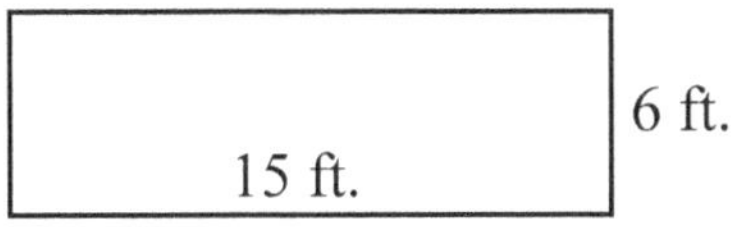

4. -7 + 9 =

Speed Drills

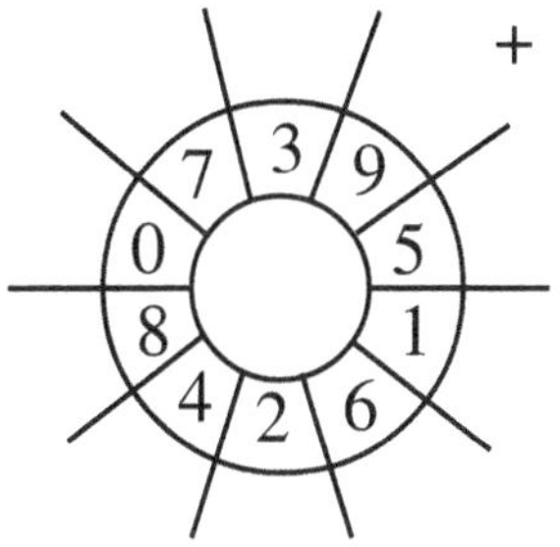

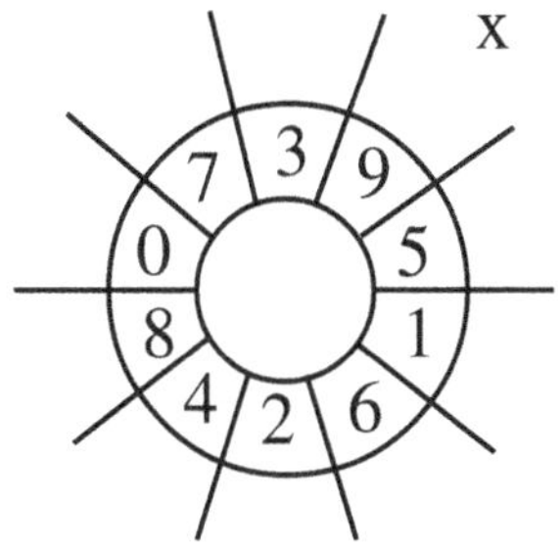

Line graphs are used to show change.

1. **Read the title.**
2. **Understand the meaning of the numbers. Estimate if necessary.**
3. **Study the data.**
4. **Answer the questions, showing work if necessary.**

HELPFUL HINTS

Use the information in the graph to answer the questions.

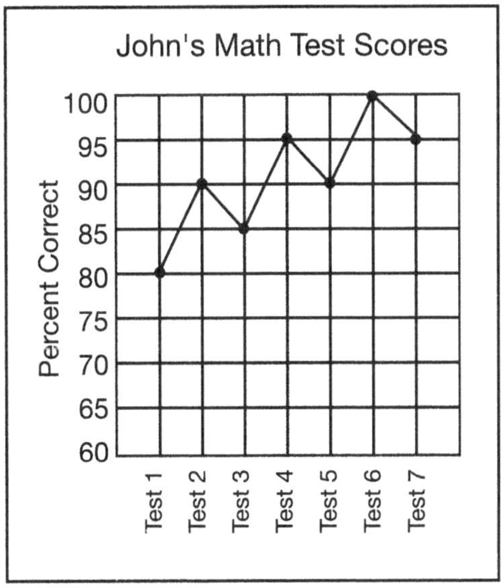

S. What was John's score on Test 5?
S. What were John's two lowest scores?
1. On which test did John score the lowest?
2. How many scores did John have that were less than 95?
3. What is the difference between his highest and lowest scores?
4. How much higher was Test 6 than Test 5?
5. How many scores did John have that were higher than 90?
6. How much higher was Test 2 than Test 1?
7. On how many tests did John have a score of 95?
8. Does John seem to be improving in math?
9. How many scores did John have that were lower than 85?
10. What was John's score on Test 3?

1	
2	
3	
4	
5	
6	
7	
8	
9	
10	
Score	

Problem Solving	A school has three fourth-grade classes, which each have 34 students. How many fourth-graders are there in the school?

It is illegal to photocopy this page. Copyright © 2023, Richard W. Fisher

Speed Drills	Review Exercises

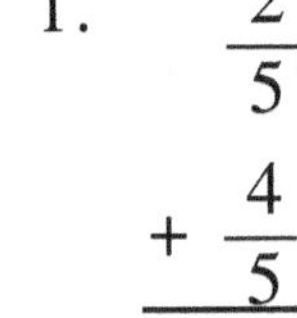

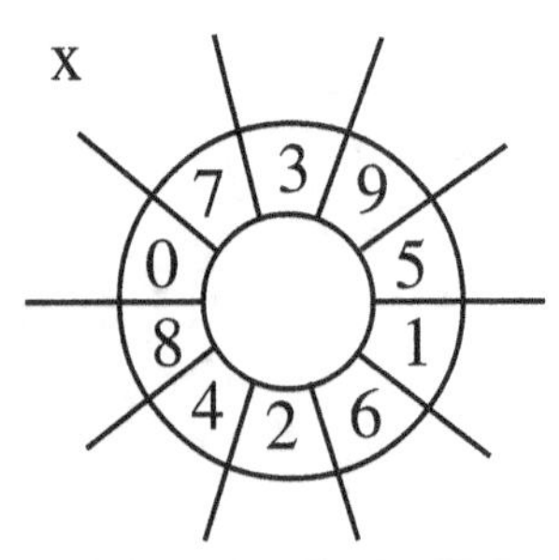

x

HELPFUL HINTS

1. $\dfrac{2}{5}$ 2. $\dfrac{7}{8}$ 3. 7 4. $4\dfrac{2}{3}$

    $+\dfrac{4}{5}$ $-\dfrac{5}{8}$ $-2\dfrac{3}{5}$ $-1\dfrac{1}{3}$

1. Read the title.
2. Understand the meaning of the numbers.
    Estimate if necessary.
3. Study the data.
4. Answer the questions, showing
    work if necessary.

| 1 |
| 2 |
| 3 |
| 4 |
| 5 |
| 6 |
| 7 |
| 8 |
| 9 |
| 10 |
| Score |

Use the information in the graph to answer the questions.

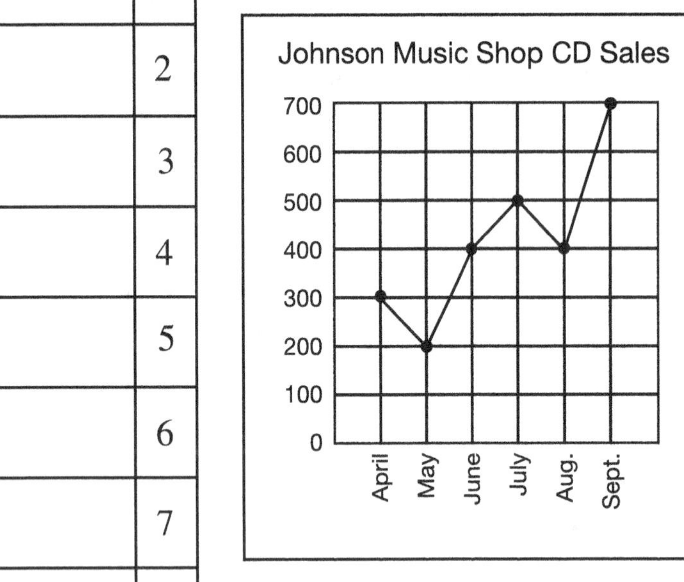

S.   In which month were the most CD's sold?

S.   How many CD's were sold in August?

1.   Which month had the lowest total sales of CD's?

2.   Which two months had the highest sales?

3.   How many more CD's were sold in July than in August?

4.   Which months had the same total sales?

5.   Between which two months was the increase in sales the greatest?

6.   What was the total number of CD's sold in the months of May and June?

7.   How many more sales were there in September than August?

8.   Which month had 300 more CD sales than May?

9.   What is the difference in sales between the month with the most sales and the month with the least sales?

10. Does the store's sales seem to be increasing or decreasing?

102   Mary's test scores were 70, 90, and 80. What was her <u>average</u> score?

Problem Solving

It is illegal to photocopy this page. Copyright © 2023, Richard W. Fisher

Review Exercises	Speed Drills

1.   12 + -18 =   2.   3 - -6 =

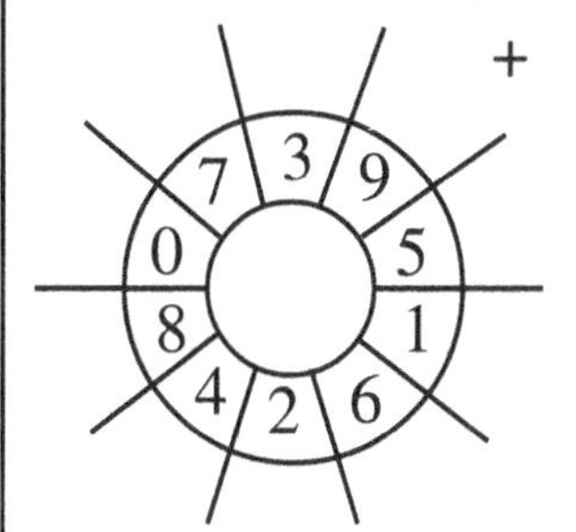

3.   4 x -6 =   4.   Find 6% of 125.

A circle graph shows the relationship between the parts to the whole and to each other.

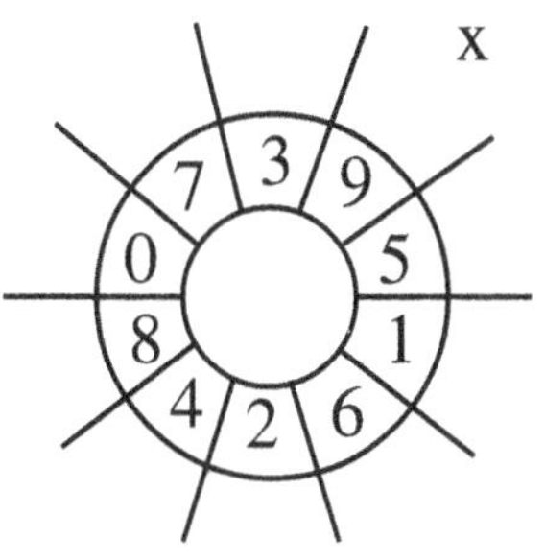

1. Read the title.
2. Understand the meaning of the numbers.
   Estimate if necessary.
3. Study the data.
4. Answer the questions, showing work if necessary.

Use the information in the graph to answer the questions.

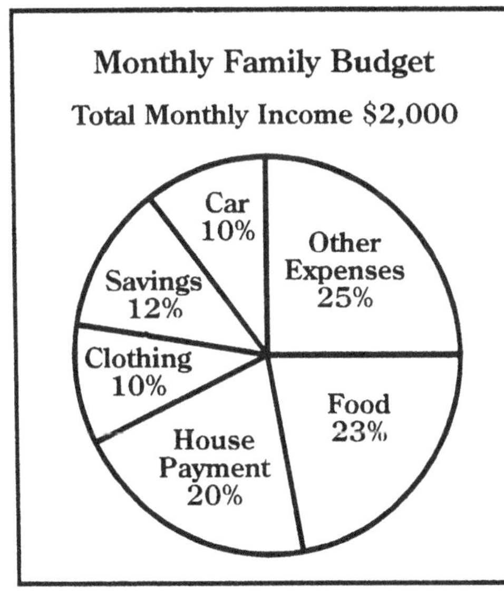

S.   What percent of the family budget is used for food?
S.   Which part of the budget requires the most expense?
1.   What percent of the budget is used to pay for food and clothing?
2.   After the car payment and house payment are paid, what percent of the budget is left?
3.   Which two items require the same part of the budget?
4.   What part of the budget would pay for medical expenses?
5.   How many dollars are spent on the house payment? (Hint: Find 20% of $2,000.00)
6.   What percent of the budget is for food and savings?
7.   What is the family's total income for the month?
8.   Which are the two most important items of the budget?
9.   What is the least important item of the budget?
10. What percent of the budget is required for cars, savings and clothing?

1	
2	
3	
4	
5	
6	
7	
8	
9	
10	
Score	

Problem Solving	A man can walk $4\frac{1}{2}$ miles in an hour. How far can he walk in 2 hours?

103

It is illegal to photocopy this page. Copyright © 2023, Richard W. Fisher

Speed Drills	Review Exercises

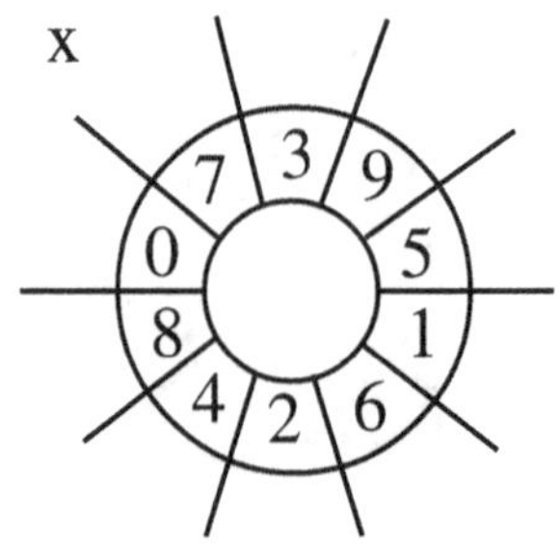

Review Exercises

1. Reduce $\frac{15}{20}$ to its lowest terms.

2. Change $\frac{1}{5}$ to a percent.

3. 2 is what % of 8?

4. $-24 \div -6 =$

1. Read the title.
2. Understand the meaning of the numbers. Estimate if necessary.
3. Study the data.
4. Answer the questions, showing work if necessary.

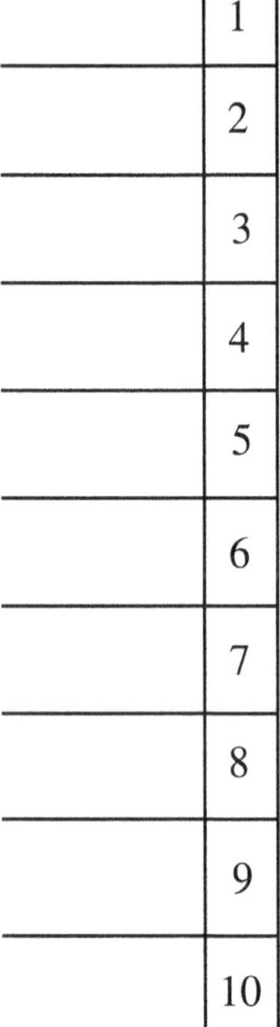

1	
2	
3	
4	
5	
6	
7	
8	
9	
10	
Score	

Use the information in the graph to answer the questions.

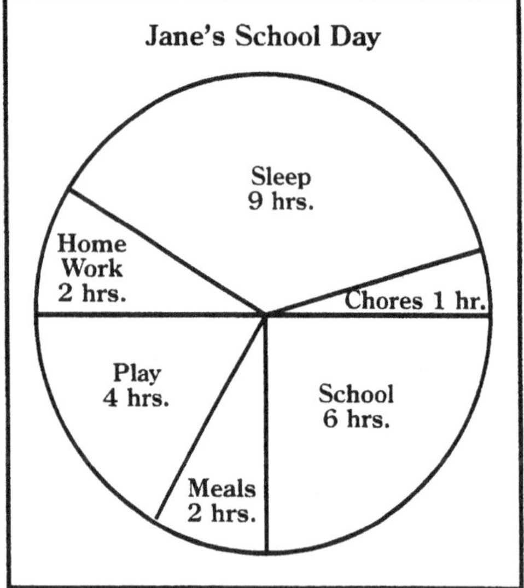

S. How many hours per day does Jane play?

S. How much time does Jane spend each day at school and doing homework?

1. If Jane starts her homework at 7:00 P.M., what time would she be finished?

2. How many more hours does Jane sleep per day than play?

3. How many hours does Jane spend at school per week?

4. If Jane goes to bed at 9:00 P.M., what time does she get up in the morning?

5. If school starts at 9:00 A.M., what time is school dismissed?

6. How many hours does Jane spend in school in 4 days?

7. How many hours does Jane spend on homework in 3 days?

8. Which three activities take up the most of her time?

9. How many more hours does Jane spend playing than doing homework each day?

10. If Jane's grades in school begin to drop, what could she do with her daily schedule to help bring them up?

A bedroom is in the shape of a rectangle with length 14 feet, and width 12 feet. What is the <u>area</u> of the bedroom?

Problem Solving

It is illegal to photocopy this page. Copyright © 2023, Richard W. Fisher

Review Exercises	Speed Drills

1. $385 + 226 + 43 =$ 2. $\begin{array}{r} 234 \\ \times\ 7 \\ \hline \end{array}$

3. $\begin{array}{r} 500 \\ -\ 236 \\ \hline \end{array}$ 4. $5\ \overline{)\ 139}$

Picture graphs are another way to compare statistics.

1. Read the title.
2. Understand the meaning of the numbers. Estimate if necessary.
3. Study the data.
4. Answer the questions, showing work if necessary.

HELPFUL HINTS

Use the information in the graph to answer the questions.

Bikes Made By Street Bike Company

1986	🚲🚲
1987	🚲🚲🚲🚲
1988	🚲🚲🚲
1989	🚲🚲🚲🚲🚲🚲
1990	🚲🚲🚲🚲🚲
1991	🚲🚲🚲🚲🚲🚲🚲🚲

Each 🚲 represents 1,000 bikes

S. How many bikes were made in 1990?

S. How many more bikes were made in 1987 than in 1986?

1. Which year produced two times as many bikes as 1988?

2. What is the total number of bikes produced in 1986, 1987, and 1988?

3. Which two years did the company make the most bikes?

4. Which two years did the company make the fewest bikes?

5. In 1987, half the bikes made were ladies' style. How many ladies' bikes were made in 1987?

6. What is the difference in bikes produced in 1991 and 1988?

7. What is the difference between the most productive and the least productive years?

8. If 1992 is expected to be double the production of 1990, how many bikes will be built in 1992?

9. What year produced half the number of bikes made in 1991?

10. 3,000 of the bikes made in 1989 were men's style bikes. How many were ladies' style bikes?

1	
2	
3	
4	
5	
6	
7	
8	
9	
10	
Score	

Problem Solving   Bill, John, and Mary together earned \$414. If they wanted to share the money equally, how much would each one receive?

105

It is illegal to photocopy this page. Copyright © 2023, Richard W. Fisher

Speed Drills	Review Exercises

+

7
3
9
5
1
6
2
4
8
0

x

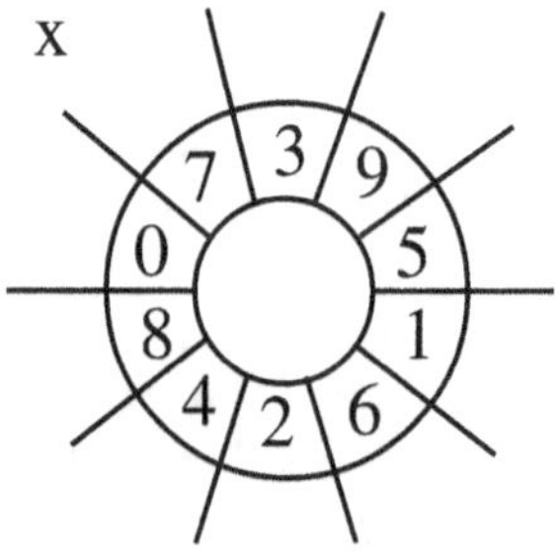
7
3
9
5
1
6
2
4
8
0

1. Find the circumference of a circle with a diameter of 6 feet.
2. Find the perimeter of a square with sides 13 feet?
3. Find the least common multiple of 5 and 3.
4. Find the greatest common factor of 15 and 20.

HELPFUL HINTS

1. Read the title.
2. Understand the meaning of the numbers. Estimate if necessary.
3. Study the data.
4. Answer the questions, showing work if necessary.

	1
	2
	3
	4
	5
	6
	7
	8
	9
	10
	Score

Use the information in the graph to answer the questions.

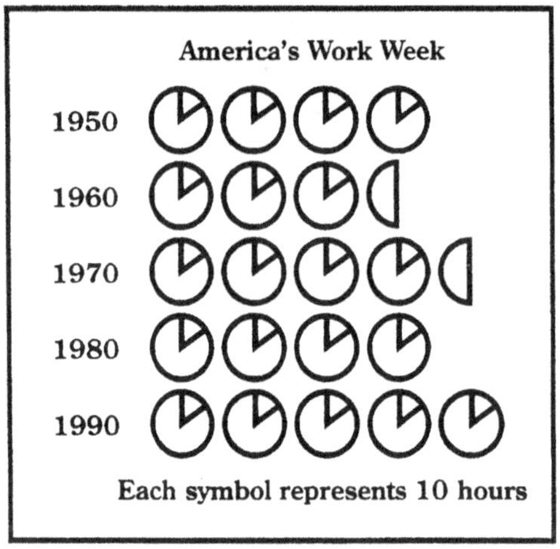
America's Work Week
1950
1960
1970
1980
1990
Each symbol represents 10 hours

S. In which year was the work week the longest?
S. How much shorter was the work week in 1960 than in 1950?
1. How many hours long was the work week in 1980?
2. How many hours long was the work week in 1960?
3. How many hours did the work week increase between 1960 and 1970?
4. Which year's work week was 45 hours?
5. How many hours less was the work week in 1960 than in 1990?
6. Any work time over 40 hours is overtime. What was the average weekly overtime in 1970?
7. Which two years had the longest work weeks?
8. If an employee worked 50 weeks in 1950, how many hours did he work altogether that year?
9. How many hours did the work week decrease between 1970 and 1980?
10. If the work week in 1991 increased to 65 hours, how much overtime would there be each week?

A team played 12 games and won 9 of them. What percent of the games did the team win? **Problem Solving**

It is illegal to photocopy this page. Copyright © 2023, Richard W. Fisher

| Review Exercises | Speed Drills |

1. $3.7 + 2.16 + 1.2 =$ 2. $7.13 - 2.67 =$

3. $\begin{array}{r} 6.5 \\ \times\ 3.2 \\ \hline \end{array}$ 4. $2\overline{)1.23}$

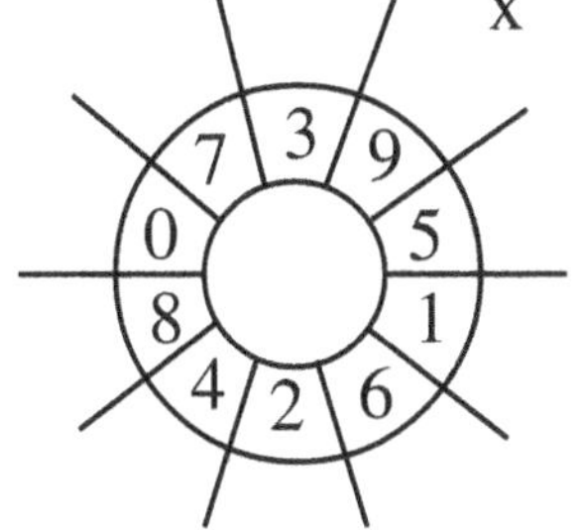

Numbers can be assigned to a point on a number line. Positive numbers are to the right of zero. Negative numbers are to the left of zero.

Examples: A is the graph of -5   A has a coordinate of -5
          B is the graph of -1   B has a coordinate of -1
          C is the graph of 5   C has a coordinate of 5

HELPFUL HINTS

Use the number line to state the coordinates of the given points:

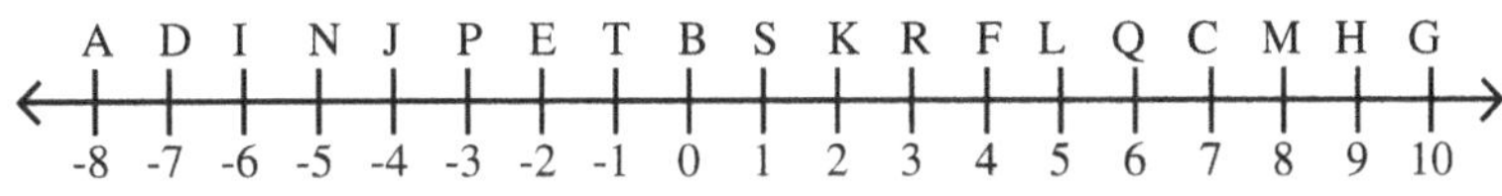

S.   B   S.   D, E, and G   1.   L and H

2.   R and F   3.   K, F, and C   4.   N and A

5.   G, H, I, and Q   6.   H, D, and S

7.   A, M, B, and P   8.   B, C, and M

9.   I, F, and P   10.   L, P, H, and A

1	
2	
3	
4	
5	
6	
7	
8	
9	
10	
Score	

| Problem Solving | A man bought 3 pens for $.35 each, and a notebook for $1.25. How much did he spend altogether? |

107

It is illegal to photocopy this page. Copyright © 2023, Richard W. Fisher

Speed Drills	Review Exercises

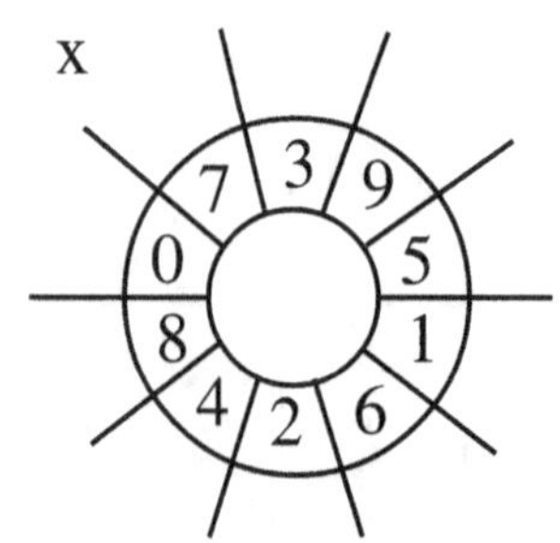

1. Find 6% of 70

2. 4 is what % of 5?

3. In a class of 40 students, 30% are boys. How many boys are in the class?

4. Sam took a test with 12 questions. If he got 9 correct, what % did he get correct?

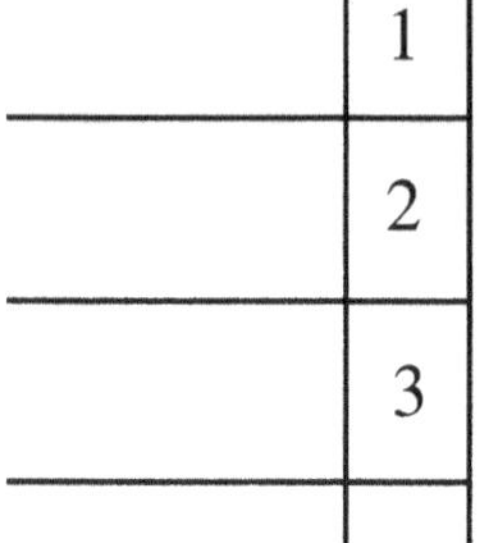

Equations can be solved and graphed on a number line.

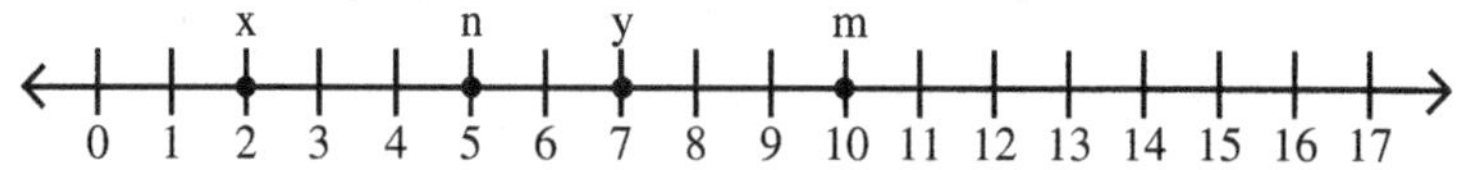

Examples:

$x + 5 = 7$ $n - 3 = 2$ $3y = 21$ $\dfrac{m}{2} = 5$

$2 + 5 = 7$ $5 - 3 = 2$ $3 \times 7 = 21$ $\dfrac{10}{2} = 5$

$\boxed{x = 2}$ $\boxed{n = 5}$ $\boxed{y = 7}$ $\boxed{m = 10}$

HELPFUL HINTS

Solve each equation and graph each solution on the number line. Also place each solution in the answer column.

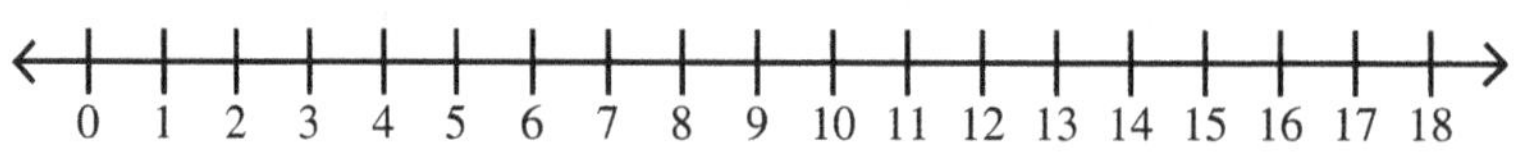

1	
2	
3	
4	
5	
6	
7	
8	
9	
10	
Score	

S. $x + 2 = 3$ S. $y - 2 = 5$ 1. $c + 4 = 7$

2. $5 - e = 0$ 3. $3d = 15$ 4. $\dfrac{f}{3} = 6$

5. $n \times 2 = 8$ 6. $3 + j = 14$ 7. $3 + k = 11$

8. $4m = 28$ 9. $6 = n + 2$ 10. $\dfrac{r}{2} = 6$

If one pencil costs $1.12, then how much would a <u>dozen</u> pencils cost?

Problem Solving

It is illegal to photocopy this page. Copyright © 2023, Richard W. Fisher

| Review Exercises | Speed Drills |

1. -3 + 6 =   2. -7 - 6 =

3. 3 x -6 =   4. -32 ÷ -2 =

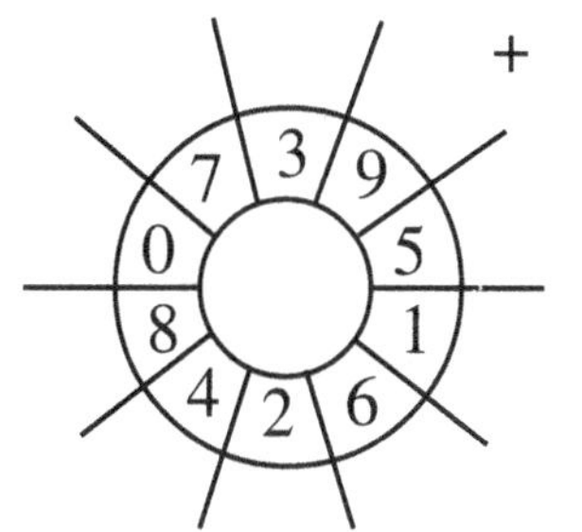

Ordered pairs can be graphed on a coordinate system.

The first number of an ordered pair shows how to move across.

The second number of an ordered pair shows how to move up and down.

Examples: To locate B, move across to the right to 3 and up 4. The ordered pair is (3,4).

To locate C, move across to the left to -5 and up 2. The ordered pair is (-5, 2).

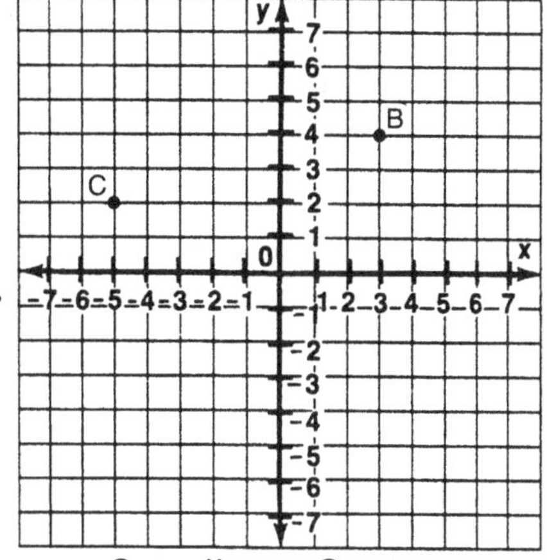

Coordinate System

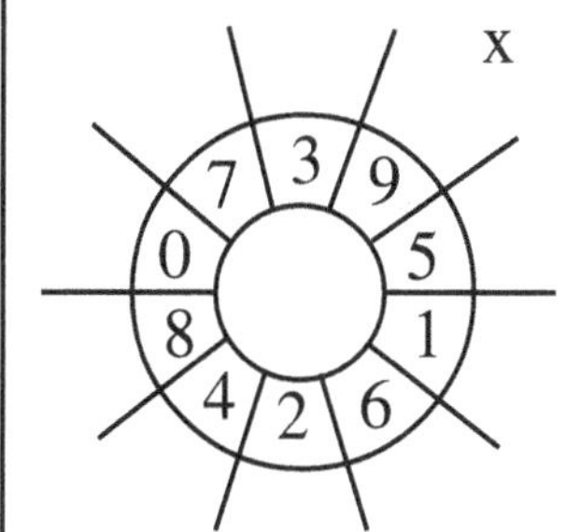

HELPFUL HINTS

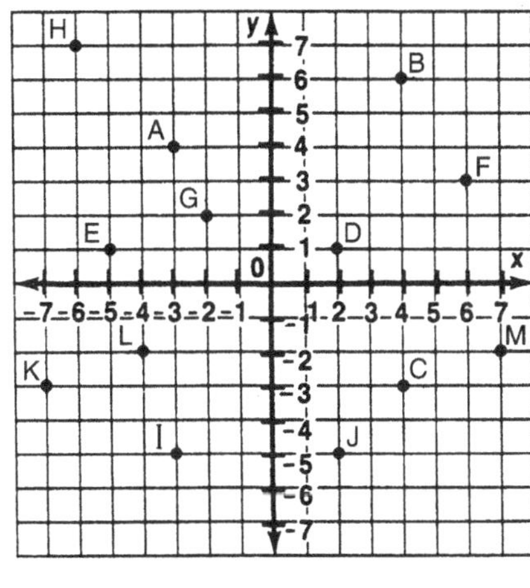

Use the coordinate system to find the ordered pair associated with each point.

S. D   S. L

1. F   2. J

3. K   4. E

5. B   6. C

7. I   8. G

9. D   10. H

1	
2	
3	
4	
5	
6	
7	
8	
9	
10	
Score	

Problem Solving | A coat was on sale for $32.25. If the regular price was $40.75, how much do you save buying the coat on sale?

109

It is illegal to photocopy this page. Copyright © 2023, Richard W. Fisher

Speed Drills	Review Exercises

Speed Drills

+

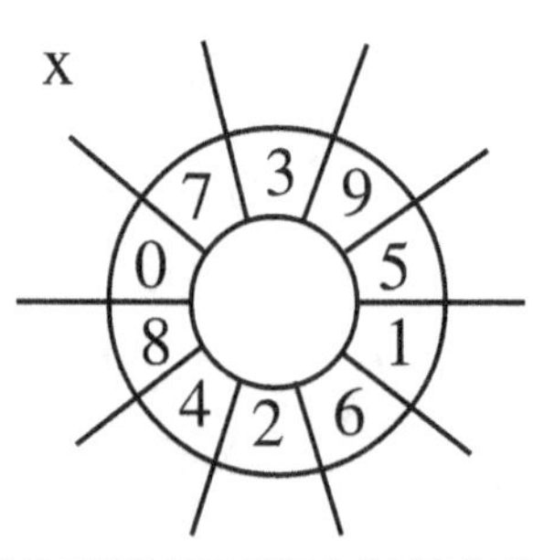

x

1.   235 + 16 + 325 =   2.   700 - 268 =

3.   524
    x 6

4.   3) 166

HELPFUL HINTS

A point can be found by matching it with an ordered pair.

Examples: (-5, 3) is found by moving across to the left to -5, and up 3. This is represented by point B.

(6, 3) is found by moving across to the right to 6, and up 3. this is represented by point C.

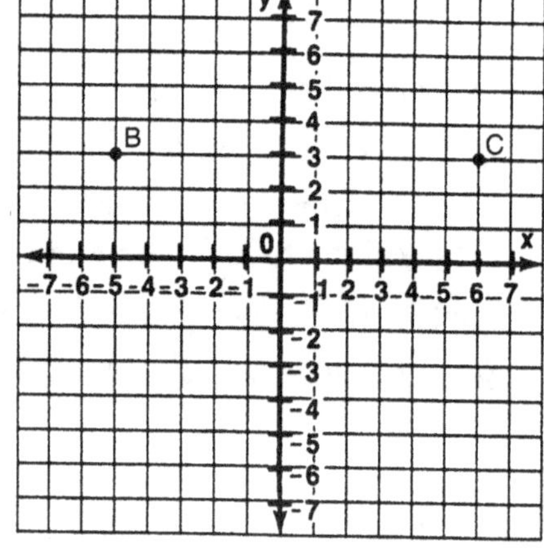

1	
2	
3	
4	
5	
6	
7	
8	
9	
10	
	Score

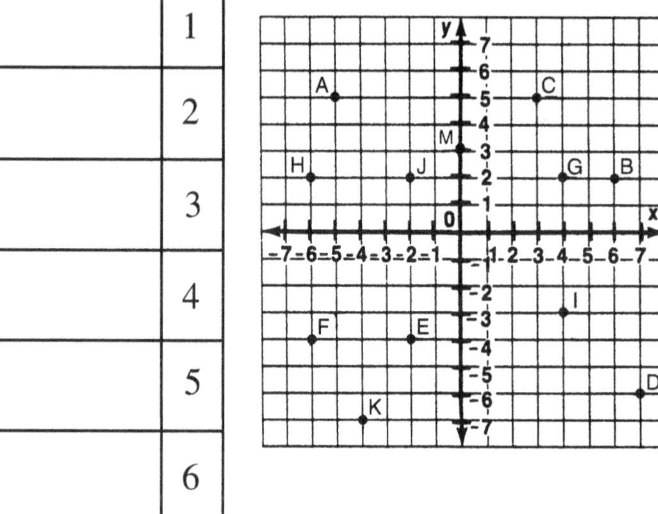

Use the coordinate system to find the point associated with each ordered pair.

S.   (6, 2)   S.   (-5, 5)

1.   (3, 5)   2.   (7, -6)

3.   (-6, -4)   4.   (0, 3)

5.   (-2, -4)   6.   (-2, 2)

7.   (-6, 2)   8.   (4, 2)

9.   (-4, -7)   10.   (4, -3)

110

If a man earns 25 dollars a day for 5 straight days, and earns 50 dollars on the sixth day, how much does he earn altogether? | **Problem Solving**

It is illegal to photocopy this page. Copyright © 2023, Richard W. Fisher

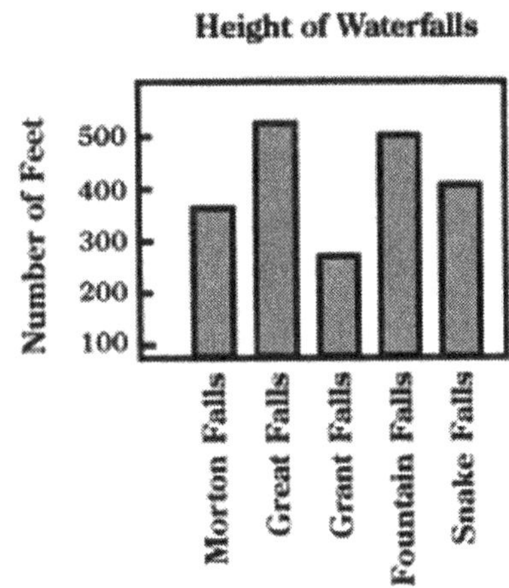

Height of Waterfalls

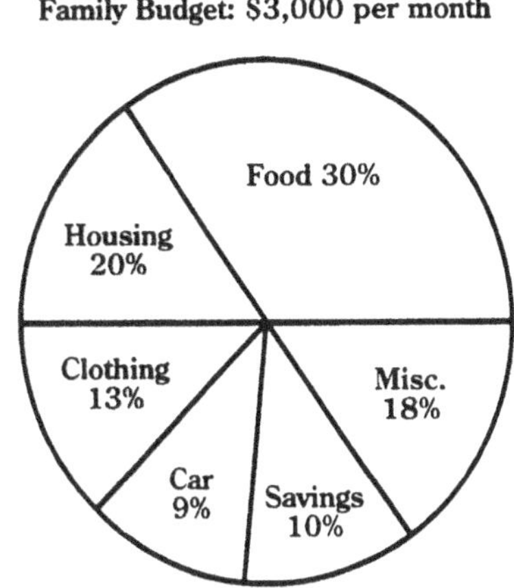

Family Budget: $3,000 per month

1. Which waterfall is the smallest?
2. Approximately how high is Great Falls?
3. Approximately how much higher is Morton Falls than Grant Falls?
4. Which waterfall is about the same height as Morton Falls?
5. Which waterfall is the fourth highest?

6. What percent of the family's money is spent on clothing?
7. What percent of the budget is spent on housing?
8. How many dollars are spent on housing per month? (Hint: Find 20% of $3,000)
9. What percent of the budget goes to savings?
10. What percent of the family budget is left after paying food, housing, and car expenses?

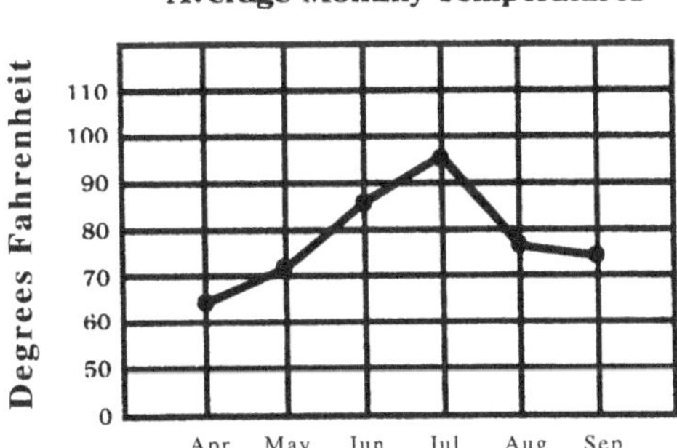

Average Monthly Temperatures

11. What is the average temperature of May?
12. How much cooler was April than July?
13. Which was the second hottest month?
14. Which month's temperature dropped the most from the previous month?
15. What is the difference in temperature between the hottest month and the second hottest month?

Fish caught in Drakes Bay in 1991

Salmon	🐟 🐟 🐟 🐟 🐟
Perch	🐟 🐟 🐟
Cod	🐟 🐟 🐟 🐟 🐟
Bass	🐟 🐟
Snapper	🐟 🐟 🐟 🐟
Tuna	🐟 🐟 🐟

Each symbol represents 100 fish

16. How many perch were caught in 1991?
17. How many more snapper were caught than bass?
18. How many salmon and cod were caught?
19. If the average perch weighs 3 pounds, how many pounds were caught in 1991?
20. What are the three most commonly caught types of fish?

1	
2	
3	
4	
5	
6	
7	
8	
9	
10	
11	
12	
13	
14	
15	
16	
17	
18	
19	
20	

111

It is illegal to photocopy this page. Copyright © 2023, Richard W. Fisher

Use the number line to state the coordinates of the given points.

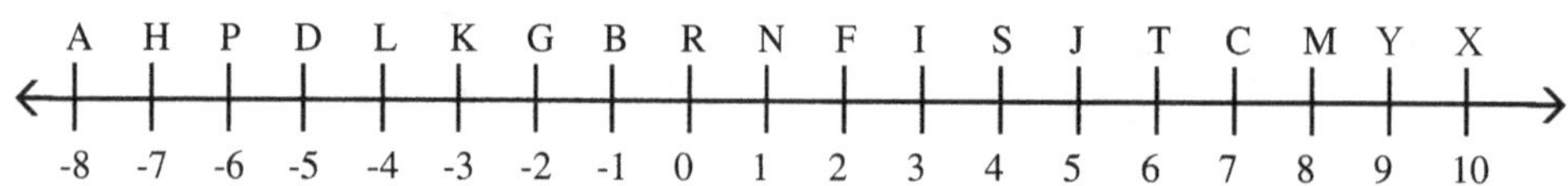

1. B   2. A, F, and G   3. G, L, and S   4. N, T, J, and H

Solve each equation and graph each solution on the number line. Be sure to label your answers. Also, place each solution in the answer column.

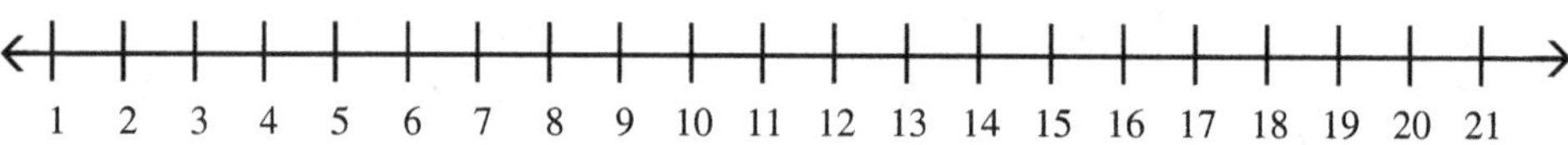

5. $n + 3 = 7$ 6. $x - 3 = 9$

7. $3y = 27$ 8. $\dfrac{m}{3} = 6$

Use the coordinate system to find the ordered pair associated with each point.

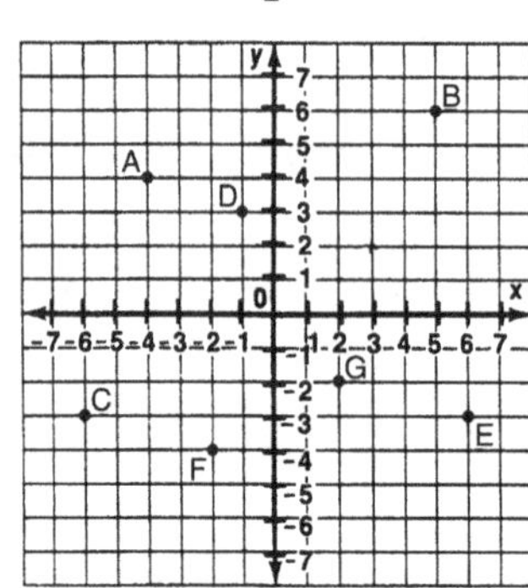

9.   A   10. E   11. B

12. F   13. C   14. G

Use the coordinate system to find the point associated with each ordered pair.

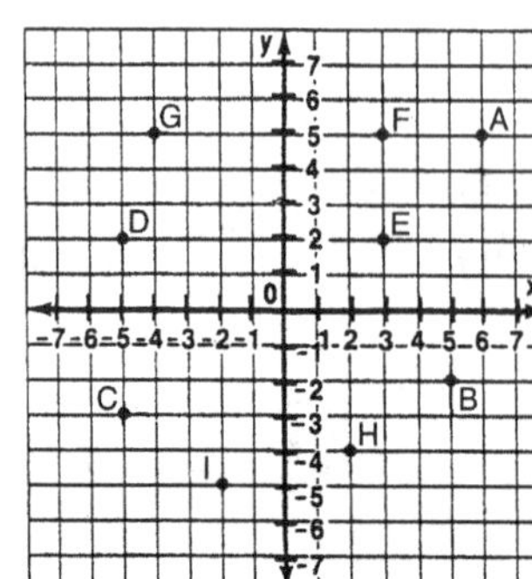

15. (5, -2)   16. (-4, 5)

17. (2, -4)   18. (-2, -5)

19. (-5, -3)   20. (3, 5)

112

It is illegal to photocopy this page. Copyright © 2023, Richard W. Fisher

Review Exercises	Speed Drills

1. $\frac{2}{3}$ $+\frac{1}{2}$

2. $\frac{7}{8}$ $-\frac{1}{8}$

3. $1\frac{1}{2} \times \frac{1}{3} =$

4. $1\frac{1}{4} \div \frac{1}{4} =$

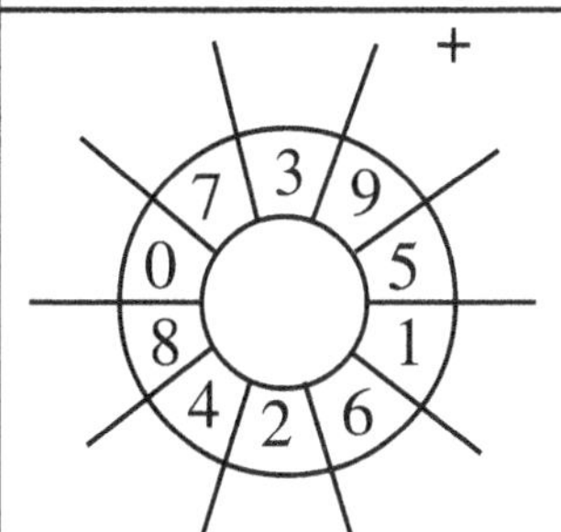

Probability tells what chance, or how likely it is for an event to occur. Probability can be written as a fraction.

$$\text{Probability} = \frac{\text{number of ways a certain outcome can occur}}{\text{number of possible outcomes}}$$

Examples: If you toss a coin, what is the probability that it will show heads?

$\frac{1}{2}$ 1 – heads is one outcome.
2 – there are two possible outcomes, heads or tails

There are 6 marbles in a jar. 3 are red, 2 are blue, and 1 is green. What is the probability that you will draw a blue one without looking?

$\frac{2}{6}$ 2 – blue marbles
6 – marbles in the jar

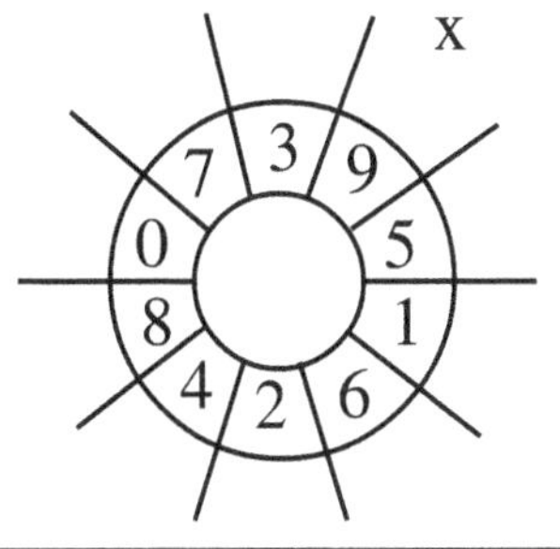

HELPFUL HINTS

Use the information below to answer the following questions.

There are 3 red marbles, 6 blue marbles, 2 black marbles, and 1 green marble in a can. Find the probability for each of the following.

S. A red marble.

S. A blue or green marble.

1. A black marble.

2. A green marble.

3. A blue or red marble.

4. Not a black marble.

5. Not a red marble.

6. Not a green or blue marble.

7. A green, red, or blue marble.

8. Not a blue marble.

9. A green, red, or black marble.

10. Not a blue or black marble.

1	
2	
3	
4	
5	
6	
7	
8	
9	
10	
Score	

Problem Solving	Adult movie tickets are \$3.50, and child tickets are \$1.25. What would it cost for two adult tickets and two child tickets?

113

It is illegal to photocopy this page. Copyright © 2023, Richard W. Fisher

Speed Drills

+

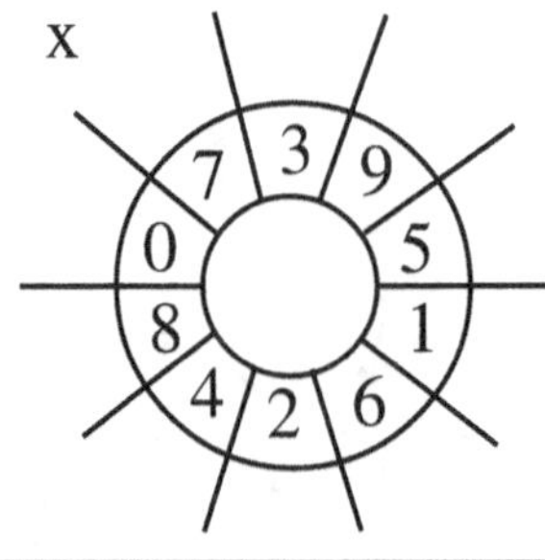

x

Review Exercises

1. $46.3 + 5 + 2.6 =$ 2. $7.2 - 3.67 =$

3. $\$3.54$ 4. $3\overline{)6.15}$
     $\times\ \ 7$

Use what you have learned to answer the following questions.

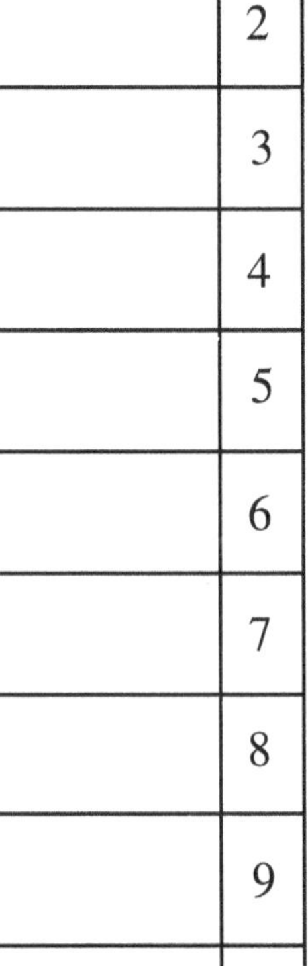

1	
2	
3	
4	
5	
6	
7	
8	
9	
10	
Score	

Using the spinner to find the probability for each of the following. Find the probability of spinning once and landing on each of the following.

S. a three   S. an even number

1. a seven   2. not a five

3. an odd number

4. a number less than five   5.   a number greater than 6

6. a nine   7. a one or an eight   8.   an even number or a 5

9. a number greater than 3   10.   a number which is a factor of 6

114

Mary has 6 dollars. Susan has three times as much money as Mary. How much money do the two girls have altogether?

Problem Solving

It is illegal to photocopy this page. Copyright © 2023, Richard W. Fisher

Review Exercises	Speed Drills

1. Change .3 to a percent

2. Change .03 to a percent.

3. Change $\frac{4}{5}$ to a percent

4. Find 5% of 65.

Statistics involves gathering and recording data. Number facts about events or objects are called data. <u>The range is the difference between the greatest number and the least number in a list of data. The mode is the number which appears the most in a list of data.</u>

Example: Find the range and mode for the list of data.
12, 10, 1, 7, 4, 7, 5
First, list numbers from least to greatest.
1, 4, 5, 7, 7, 10, 12   The range is 12 - 1 = (11)
The mode is (7) which appears the most.

HELPFUL HINTS

Arrange the data in order from least to greatest. Then find the range and mode.

S.   7, 4, 1, 8, 2, 5, 4

S.   6, 2, 7, 6, 8, 2, 5, 6, 3

1.   7, 4, 8, 2, 4, 7, 7

2.   25, 17, 30, 39, 16, 24, 30

3.   1, 3, 6, 3, 4, 6, 11, 9

4.   1, 6, 17, 8, 9, 20, 9

5.   7, 3, 1, 3, 1, 3, 8, 4

6.   3, 14, 8, 6, 11, 8, 14, 8

7.   1, 10, 2, 9, 3, 8, 2, 7

8.   85, 91, 90, 86, 91, 87

9.   1, 10, 2, 9, 2, 7, 2, 8

10.   20, 2, 19, 1, 2, 16, 3

1	
2	
3	
4	
5	
6	
7	
8	
9	
10	
Score	

Problem Solving Bill has 12 dollars. Jose has half as much money as Bill. How much money do they have altogether?

115

It is illegal to photocopy this page. Copyright © 2023, Richard W. Fisher

Speed Drills	Review Exercises

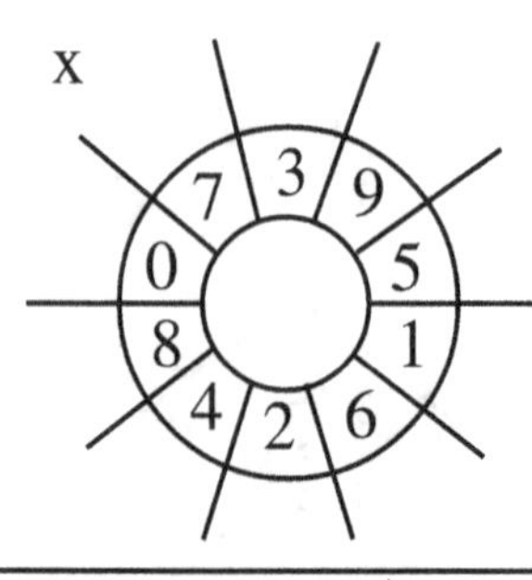

1.   -7 + 12 =   2.   -6 - 4 =

3.   3 x -8 =   4.   -24 ÷ -3 =

The mean of a list of data is found by adding all the items in the list and then dividing by the number of items.

The median is the middle number, when the list of data is arranged from least to greatest.

Example: Find the mean and median for the list of data.
            1, 2, 5, 6, 6

Median = (5)   Mean = $\dfrac{1 + 2 + 5 + 6 + 6}{5} = \dfrac{20}{5} = (4)$

HELPFUL HINTS

Arrange the data in order from least to greatest. Then find the mean and median for each list of data.

S.   1, 5, 2, 4, 3

S.   6, 1, 7, 4, 2, 6, 2

1.   2, 7, 1, 4, 1

2.   1, 5, 7, 1, 2, 2, 3

3.   5, 25, 10, 20, 15

4.   1, 1, 1, 3, 3, 3, 4, 1, 1

5.   8, 5, 2, 9, 3, 6, 9

6.   126, 136, 110

7.   7, 3, 4, 2, 4

8.   3, 1, 4, 7, 5

9.   2, 10, 4, 8, 1

10.   50, 70, 30

1	
2	
3	
4	
5	
6	
7	
8	
9	
10	
Score	

If 3 cans of juice cost $1.14, what is the cost of 1 can?   **Problem Solving**

It is illegal to photocopy this page. Copyright © 2023, Richard W. Fisher

Review Exercises	Speed Drills

1. 6 ft. rectangle, 3 ft.

Perimeter =

2. 13 ft. square

Area =

3. 3 ft. circle

Circumference =

4. rectangle 12 ft., 6 ft.

Perimeter =

Use what you have learned to answer the following questions.

HELPFUL HINTS

Arrange the data in order from least to greatest than answer the questions.

2, 8, 6, 2, 7

S. What is the range?   S. What is the mode?

1. What is the mean?   2. What is the median?

1, 9, 2, 7, 2, 3, 4

3. What is the median?   4. What is the mode?

5. What is the range?   6. What is the mean?

2, 11, 8, 6, 1, 2, 5

7. What is the range?   8. What is the mode?

9. What is the mean?   10. What is the median?

1	
2	
3	
4	
5	
6	
7	
8	
9	
10	
Score	

Problem Solving Alex had test scores of 80, 70, and 90. What was his mean score?

It is illegal to photocopy this page. Copyright © 2023, Richard W. Fisher

There are 4 green marbles, 3 red marbles, 2 white marbles, and 1 blue marble in a can. What is the probability for each of the following?

1. a red marble   2. a green marble   3. a green or blue marble

4. not a red marble   5. a green, red or blue marble   6. not a green marble

Use the spinner to find the probability of spinning once and landing on each of the following.

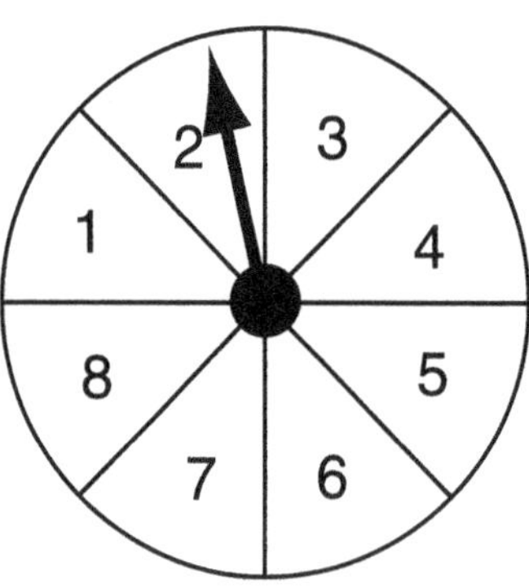

7. a five   8. an odd number

9. a number greater than three

10. a one or a three

11. a number less than five

12. a one or a six

Arrange the data in order from least to greatest, then answer the questions.

2, 5, 4, 10, 4

13.   What is the range?   14.   What is the mode?

15.   What is the mean?   16.   What is the median?

2, 5, 2, 1, 3, 7, 8

17.   What is the mode?   18.   What is the mean?

19.   What is the range?   20.   What is the median?

118

It is illegal to photocopy this page. Copyright © 2023, Richard W. Fisher

Review Exercises	Speed Drills

1. Express as a ratio.

2. Solve the equation
$$x - 7 = 10$$

3. Solve the equation.
$$x + 4 = 12$$

4. Solve the equation.
$$\frac{x}{3} = 5$$

1. Read the problem carefully. *Sometimes drawing a practice diagram can help.*
2. Find the important facts and numbers. *Sometimes a formula is necessary.*
3. Decide what operations to use. *Sometimes reading a problem more than once helps.*
4. Solve the problem. *Show your work.*

HELPFUL HINTS

S. There are 2 fourth grade classes with enrollment of 32 and 35. How many fourth graders are there in all?

S. A family drove 215 miles each day for 5 days. How far did they drive altogether?

1. The attendance at the Raider's game last year was 1,235. This year 1,783 attended. What was the increase in attendance?

2. 5 friends earned 735 dollars. If they want to divide the money evenly, how much will each person receive?

3. A school put on a play. Monday 265 attended, 136 attended Tuesday and 243 attended Wednesday. What was the total attendance?

4. John scored 1,234 points on a video game. Julie scored 1,455 points. How many more points did Julie receive than John?

5. If a car travels 65 miles per hour, how far does it travel in 8 hours?

6. A car traveled 240 miles in 4 hours. What was the average speed per hour?

7. A car's gas tank holds 8 gallons. If the car can travel 13 miles per gallon, how far can the car travel?

8. A banner is in the shape of a triangle with sides 14 feet, 15 feet, and 17 feet. How many feet is it around the banner?

9. A rope is 112 feet long. If it is cut into pieces 2 feet long, how many pieces will there be?

10. A library has 2,160 fiction books, 1,160 nonfiction books, and 253 reference books. How many books are there in all?

Speed Drills	
1	
2	
3	
4	
5	
6	
7	
8	
9	
10	
Score	

119

It is illegal to photocopy this page. Copyright © 2023, Richard W. Fisher

Speed Drills	Review Exercises

+

(speed drill wheel: 7 3 9 0 5 8 1 4 2 6)

x

(speed drill wheel: 7 3 9 0 5 8 1 4 2 6)

1. $\dfrac{3}{5}$
    $+\ \dfrac{3}{5}$

2. $\dfrac{7}{8}$
    $-\ \dfrac{1}{4}$

3. $\dfrac{2}{3}$ of $\dfrac{3}{4}$ =

4. $\dfrac{1}{2}$ of 8 =

HELPFUL HINTS

1. Read the problem carefully.
2. Find the important facts and numbers.
3. Decide what operations to use.
4. Solve the problem.

*Sometimes drawing a practice diagram can help.
*Sometimes a formula is necessary.
*Sometimes reading a problem more than once helps.
*Show your work.

1	
2	
3	
4	
5	
6	
7	
8	
9	
10	
Score	

S. A cook uses $2\frac{1}{5}$ cups of flour for a cake and $1\frac{1}{3}$ cups of flour for a pie. How much flour did he use?

1. Bill weighs $90\frac{1}{2}$ pounds. Sam weighs $93\frac{3}{4}$ pounds. How much more does Sam weigh than Bill?

3. A book weighs $1\frac{1}{2}$ pounds. How many pounds do 3 books weigh?

5. A flag is in the shape of a square. Each side is $3\frac{1}{2}$ feet long. What is the perimeter of the flag?

7. A man worked $2\frac{1}{2}$ hours on Monday and $1\frac{1}{4}$ hours on Tuesday. How many hours did he work altogether?

9. A girl can jog 6 miles in an hour. At this pace how far can she jog in $2\frac{1}{2}$ hours?

S. A ribbon is $2\frac{1}{2}$ feet long. It was cut into pieces $\frac{1}{2}$ feet long. How many pieces were there?

2. Steve earned 15 dollars and spent $\frac{2}{3}$ of it. How much did he spend?

4. A woman can drive to work in $7\frac{1}{2}$ minutes. She can drive home in $5\frac{1}{2}$ minutes. What is the total time she has to drive to and from work?

6. Natalie earned $5\frac{1}{2}$ dollars and spent $2\frac{1}{4}$ dollars. How much money did she have left?

8. Linda worked $3\frac{1}{2}$ hours on Monday and $1\frac{3}{5}$ hours on Tuesday. How many more hours did she work on Monday than on Tuesday?

10. A factory can make a tire in $1\frac{1}{2}$ minutes. How many tires can it make in 6 minutes?

120

It is illegal to photocopy this page. Copyright © 2023, Richard W. Fisher

| Review Exercises | Speed Drills |

1. $1.23 + $3.16 + $4.24 =   2. $7.00
                                                + $2.58

3. $2.17   4.   6) 4.14
     x   6

1. Read the problem carefully.   *Sometimes drawing a practice diagram can help.
2. Find the important facts and numbers. *Sometimes a formula is necessary.
3. Decide what operations to use.   *Sometimes reading a problem more than once helps.
4. Solve the problem.   *Show your work.

HELPFUL HINTS

S. If a dozen pencils cost $1.17, how much do 6 dozen cost?

S. A desk is shaped like a rectangle. If the length is 2.6 feet, and the width is 1.3 feet, what is the perimeter of the desk?

1. Beef costs $3.15 per pound. Fish costs $2.85 per pound. How much more is beef per pound.

2. A man bought a chair for $35.50, a desk for $125.55, and a lamp for $24.75. What was the total cost of the items?

3. 5 cans of pet food cost $7.25. What is the price per can?

4. A car travels 50.6 miles in one hour. At this rate, how far can it travel in 5 hours?

5. Steve can earn $7.25 per day. How much can he earn in 5 days?

6. A sack of potatoes costs $2.50. If there are 5 pounds of potatoes in a sack, what is the price per pound?

7. A bike was on sale for $125.50. If the regular price was $175.75, how much could he save buying it on sale?

8. Ellen weighs 85.2 pounds. Mary weighs 75.9 pounds. What is their combined weight?

9. If 3 pounds of butter cost $5.34, what is the price per pound?

10. A man bought a quart of milk for $1.23. If he paid with a $5.00 bill, how much change should he get?

Speed Drills	
1	
2	
3	
4	
5	
6	
7	
8	
9	
10	
Score	

121

It is illegal to photocopy this page. Copyright © 2023, Richard W. Fisher

Speed Drills	Review Exercises

+

x

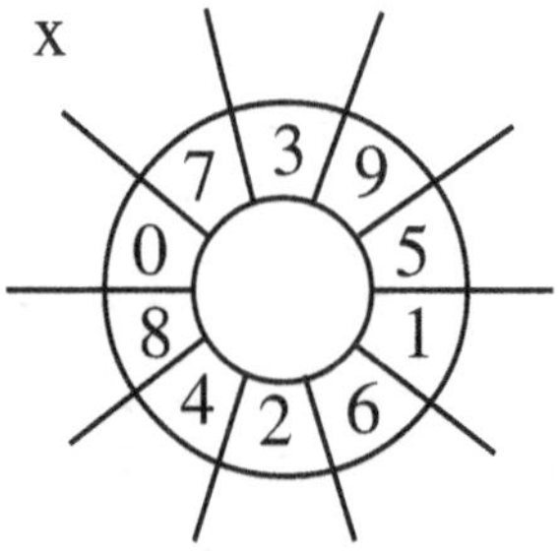

1. $\frac{1}{2} + \frac{1}{3}$

2. $3.4 + 2.66 =$

3. $\frac{7}{8} - \frac{1}{2}$

4. $3.7 - 1.63 =$

HELPFUL HINTS

1. Read the problem carefully.
2. Find the important facts and numbers.
3. Decide what operations to use.
4. Solve the problem.

*Sometimes drawing a practice diagram can help.
*Sometimes a formula is necessary.
*Sometimes reading a problem more than once helps.
*Show your work.

	1
	2
	3
	4
	5
	6
	7
	8
	9
	10
	Score

S. A rope 64 feet long is to be cut into pieces 4 feet long. How many pieces will there be?

S. If candy bars cost $.15 each, how much will 6 candy bars cost?

1. Tim had $3.25. If he spent $1.55, how much money does he have left?

2. 4 loaves of bread cost $4.24. How much does 1 loaf of bread cost?

3. Steve had a total score of 285 on 3 tests. What was his average score?

4. A boy can earn $1\frac{1}{2}$ dollars in 1 hour. How much can he earn in 4 hours?

5. A carpenter needs to glue together two boards with lengths of 3.8 inches and 7.6 inches. What would the length be after they are glued together?

6. A school has 600 students. If they are to be grouped into 20 equally sized classes, how many students will be in each class?

7. A rancher owns 40 cows. If he decides to sell $\frac{1}{5}$ of them, how many cows would he sell?

8. A farm is in the shape of a square. If each side is 7 miles, what is the perimeter of the farm?

9. A board was 7 feet long. If $2\frac{1}{2}$ feet were cut off, how much of the board was left?

10. Steak costs $2.70 per pound. How much will 5 pounds of steak cost?

It is illegal to photocopy this page. Copyright © 2023, Richard W. Fisher

Final Review – Whole Numbers

1	
2	
3	
4	
5	
6	
7	
8	
9	
10	
11	
12	
13	
14	
15	
16	
17	
18	
19	
20	

1. $\begin{array}{r} 34 \\ +\ 23 \\ \hline \end{array}$ 2. $\begin{array}{r} 356 \\ +\ 374 \\ \hline \end{array}$ 3. $\begin{array}{r} 26 \\ 23 \\ +\ 34 \\ \hline \end{array}$ 4. $\begin{array}{r} 346 \\ +\ 524 \\ \hline \end{array}$

5. $328 + 26 + 145 =$ 6. $\begin{array}{r} .243 \\ +\ .327 \\ \hline \end{array}$

7. $\begin{array}{r} 27 \\ -\ 15 \\ \hline \end{array}$ 8. $\begin{array}{r} 342 \\ -\ 125 \\ \hline \end{array}$ 9. $\begin{array}{r} 90 \\ -\ 76 \\ \hline \end{array}$

10. $\begin{array}{r} 713 \\ -\ 245 \\ \hline \end{array}$ 11. $\begin{array}{r} 542 \\ -\ 263 \\ \hline \end{array}$

12. $\begin{array}{r} 22 \\ \times\ 3 \\ \hline \end{array}$ 13. $\begin{array}{r} 27 \\ \times\ 4 \\ \hline \end{array}$ 14. $\begin{array}{r} 523 \\ \times\ 4 \\ \hline \end{array}$

15. $\begin{array}{r} 35 \\ \times\ 24 \\ \hline \end{array}$ 16. $\begin{array}{r} 205 \\ \times\ 4 \\ \hline \end{array}$ 17. $2\overline{)46}$

18. $4\overline{)526}$ 19. $4\overline{)139}$ 20. $22\overline{)4664}$

It is illegal to photocopy this page. Copyright © 2023, Richard W. Fisher

Final Review – Fractions and Mixed Numerals

1. $\dfrac{1}{5} + \dfrac{2}{5}$

2. $\dfrac{3}{7} + \dfrac{5}{7}$

3. $\dfrac{2}{5} + \dfrac{1}{2}$

4. $\dfrac{1}{3} + \dfrac{1}{2}$

5. $3\frac{1}{3} + 2\frac{1}{2}$

6. $\dfrac{3}{5} - \dfrac{1}{5}$

7. $5 - 2\frac{1}{3}$

8. $\dfrac{3}{4} - \dfrac{1}{2}$

9. $3\frac{2}{3} - 1\frac{1}{2}$

10. $5\frac{3}{5} - 1\frac{1}{2}$

11. $\dfrac{1}{2} \times \dfrac{1}{4} =$

12. $\dfrac{3}{4} \times \dfrac{5}{6} =$

13. $\dfrac{1}{3} \times 9 =$

14. $2 \times \dfrac{3}{4} =$

15. $\dfrac{1}{2} \times 5 =$

16. $1\frac{1}{2} \times 2 =$

17. $\dfrac{1}{3} \div \dfrac{1}{2} =$

18. $\dfrac{2}{3} \div \dfrac{1}{2} =$

19. $1\frac{1}{2} \div \dfrac{1}{2} =$

20. $2\frac{1}{2} \div \dfrac{1}{4} =$

124

It is illegal to photocopy this page. Copyright © 2023, Richard W. Fisher

Final Review – Decimal Operations

1.   3.2
   + 4.3

2.   3.12
   + 2.3

3.   4.3 + 3.14 =

4. $2.62
  + $3.34

5.   3.16 + 2 + 4.16 =

6.   5.1
   − 3.2

7.   3.2
   − 1.14

8.   6.42
   − 1.23

9.   5.1
   − 2.36

10. 5.3 − 1.6 =

11. 2.1
   x 5

12. 1.24
   x 3

13. .413
   x 5

14. .23
   x 21

15. 2.14
   x 3

16. 2.13
   x 5

17. 2) 4.6

18. 3) .123

19. 2) .23

20. 2) 3.48

It is illegal to photocopy this page. Copyright © 2023, Richard W. Fisher

Final Review – Ratio, Proportion, Percent

Write numbers 1 and 2 as a ratio expressed in fraction form.

1. 5 pennies to 2 dimes   2. 7 to 6

For numbers 3 and 4 solve each proportion.

3. $\dfrac{2}{3} = \dfrac{4}{?}$ 4. $\dfrac{3}{4} = \dfrac{6}{?}$

Change numbers 5 through 8 to a percent.

5. $\dfrac{17}{100} =$ 6. $\dfrac{3}{10} =$ 7. $.13 =$ 8. $.4 =$

Change numbers 9 through 11 to a decimal and a fraction.

9. $12\% = .\quad = \quad -$ 10. $3\% = .\quad = \quad -$

11. $20\% = .\quad = \quad -$

12. Find 12% of 30.   13. Find 3% of 60.

14. 2 is what % of 5?   15. 2 is what % of 8?

16. Change $\dfrac{1}{5}$ to a percent. 17. Change $\dfrac{1}{4}$ to a percent.

18. A man earned 30 dollars. If he put 20% of it into the bank, how much did he put into the bank?

19. A girl baked 8 cakes. If she sold 2 of them, what percent did she sell?

20. Al took a test with 20 questions and got 20% correct. How many questions did he get correct?

126

It is illegal to photocopy this page. Copyright © 2023, Richard W. Fisher

Final Review – Geometry

Use this figure to answer questions 1-8

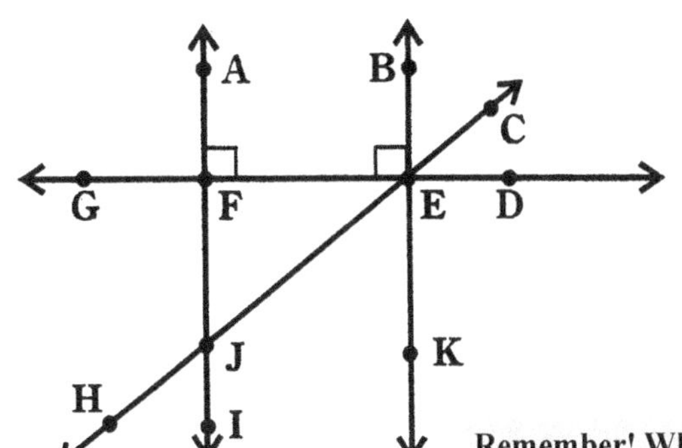

1. Name 2 parallel lines
2. Name 2 perpendicular lines
3. Name a line segment
4. Name a ray
5. Name an acute angle
6. Name an obtuse angle
7. Name a straight angle
8. Name a right angle

Remember! When naming an angle, the letter in the middle is the vertex.

Triangle A

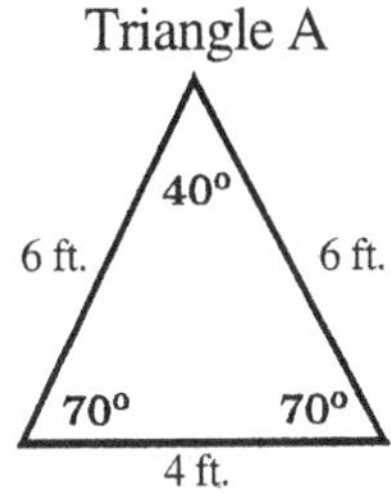

Triangle B

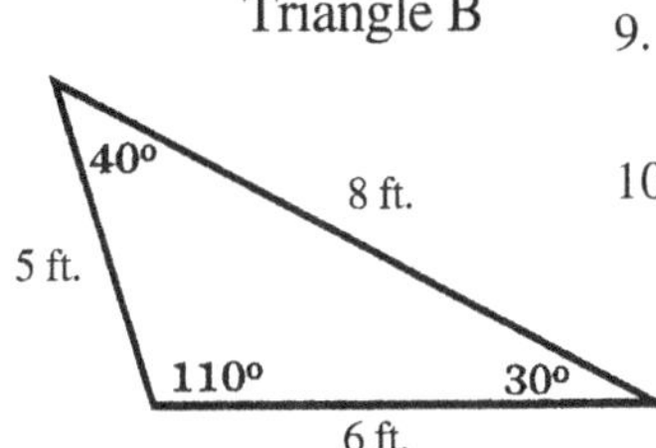

9. Classify Triangle A by its sides and angles
10. Classify Triangle B by its sides and angles

11. Find the perimeter

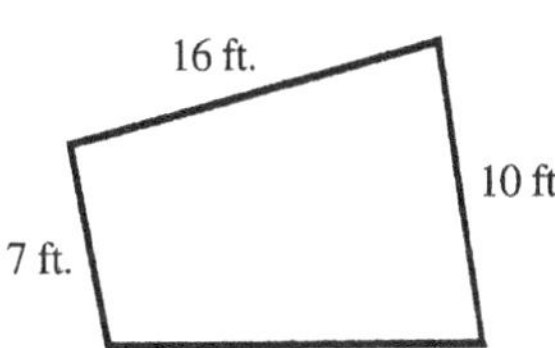

12. Find the Circumference

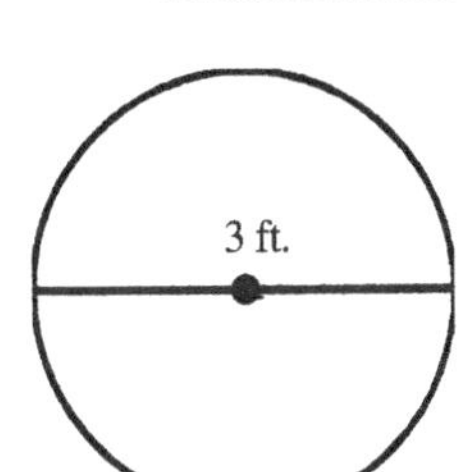

13. What is the length of the diameter?

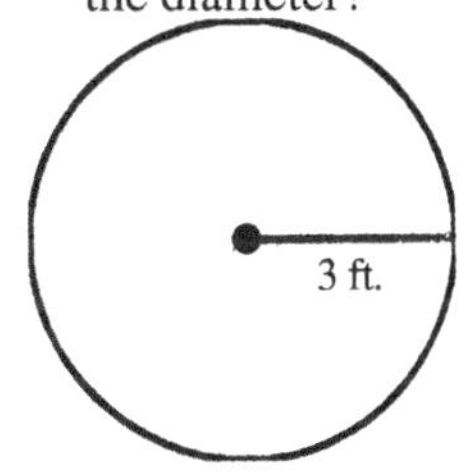

14. Find the area.

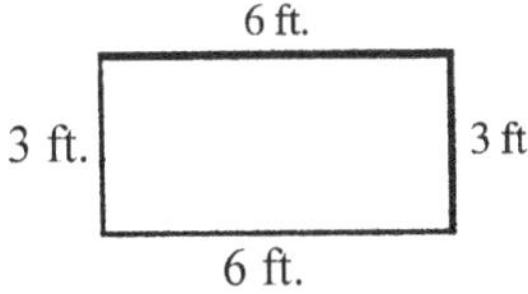

15. Find the perimeter.

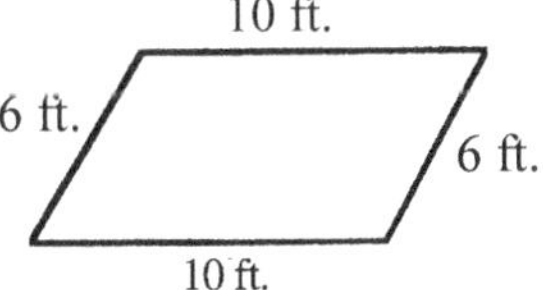

16. Find the perimeter.

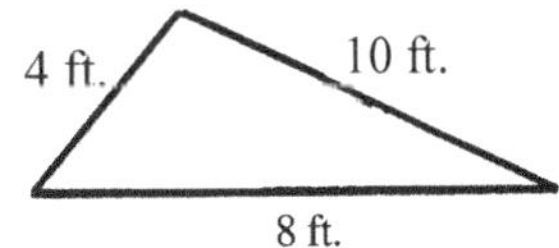

17. Find the area.

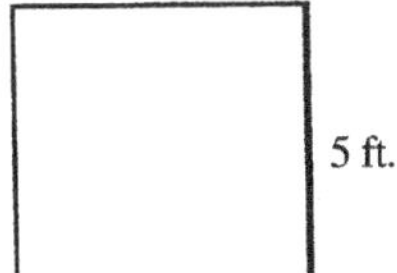

18. Identify and count the parts.

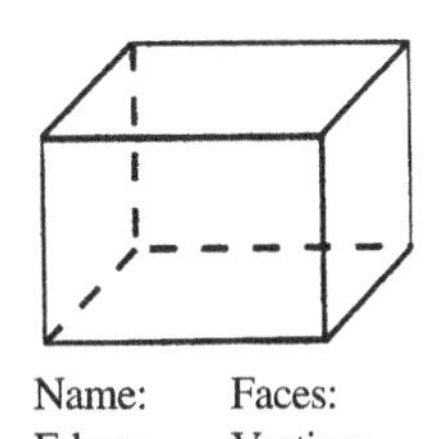

Name: Faces:
Edges   Vertices

19. Find the perimeter.

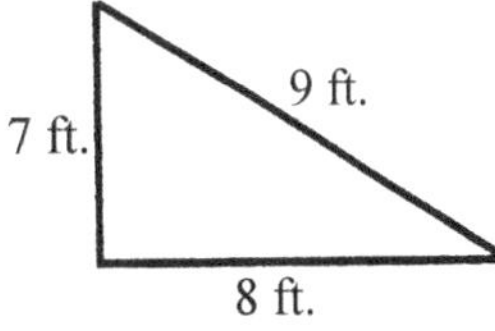

20. Find the perimeter of a square with sides of 16 ft.

127

It is illegal to photocopy this page. Copyright © 2023, Richard W. Fisher

#	
1	
2	
3	
4	
5	
6	
7	
8	
9	
10	
11	
12	
13	
14	
15	
16	
17	
18	
19	
20	

Final Review – Number Theory and Algebra

Find all factors of each number.

1. 12   2. 15   3. 20

Find the greatest common factor of each pair of numbers.

4. 6 and 8   5. 8 and 12   6. 20 and 15

Complete the list of multiples of each number.

7. 2: 0, 2, 4, ☐, ☐, ☐

8. 3: 0, ☐, 6, ☐, ☐, 15

9. 4: ☐, ☐, ☐, ☐, ☐, 20

Find the least common multiples of each of the pair of numbers.

10. 3 and 2   11. 4 and 6   12. 4 and 8

Solve each of the following equations.

13. $x + 1 = 3$ 14. $x - 2 = 3$ 15. $3x = 15$

16. $\dfrac{x}{2} = 4$ 17. $7 = x + 4$ 18. $5 - n = 2$

19. $x \div 2 = 4$ 20. $3n = 12$

128

It is illegal to photocopy this page. Copyright © 2023, Richard W. Fisher

1.   5 + -3 =

2.   -5 + 3 =

3.   -5 + -3 =

4.   -6 + -2 =

5.   7 + -5 =

6.   -3 + -5 =

7.   2 - 6 =

8.   2 - -5 =

9.   -2 - 3 =

10.   3 x -5 =

11.   -3 x -6 =

12.   -6 x 4 =

13.   -8 x -4 =

14.   3 x -12 =

15.   -6 x -4 =

16.   8 ÷ -2 =

17.   -12 ÷ 3 =

18.   -14 ÷ -2 =

19. $\dfrac{-12}{2} =$

20. $\dfrac{-9}{-3} =$

1	
2	
3	
4	
5	
6	
7	
8	
9	
10	
11	
12	
13	
14	
15	
16	
17	
18	
19	
20	

It is illegal to photocopy this page. Copyright © 2023, Richard W. Fisher

Final Review – Charts and Graphs

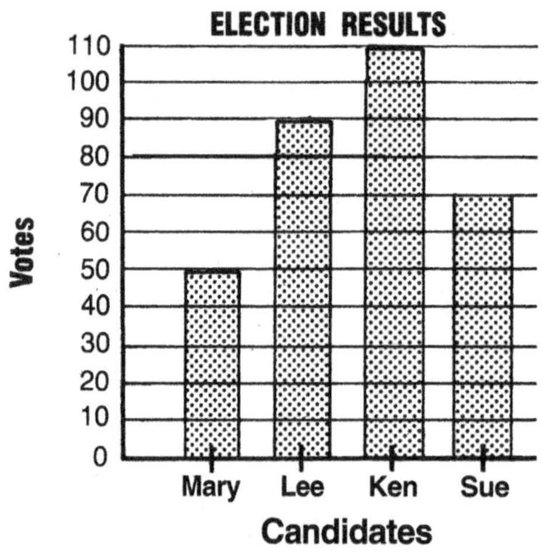

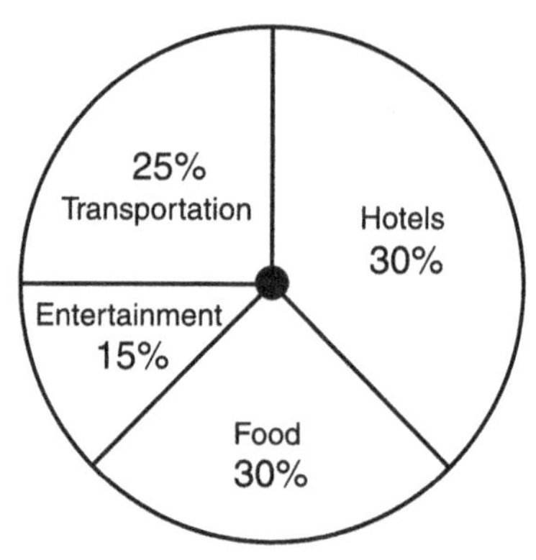

1. Which candidate got the most votes?
2. How many votes did Sue receive?
3. How many more votes did Lee get than Mary?
4. Together, how many votes did Lee and Ken receive?
5. Who got the third most votes?

6. What percent is spent on hotels?
7. What percent of the budget is spent on transportation?
8. What percent is spent altogether on transportation and hotels?
9. What is the second largest part of the budget?
10. What percent of the budget is used for entertainment?

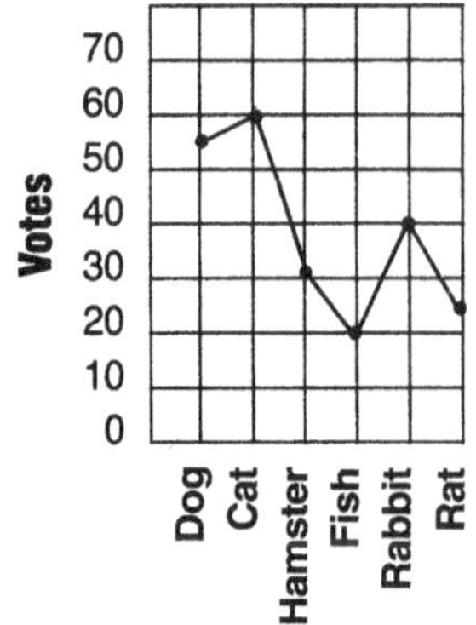

11. How many picked cats as their favorite animal?
12. What was the favorite animal?
13. What was the least favorite animal?
14. How many more people picked cats than hamsters?
15. What was the second favorite animal?

Cars sold at Atlas Car Company

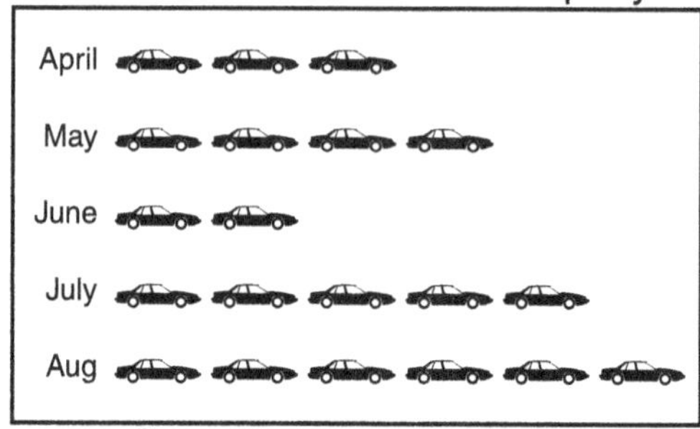

Each symbol represents 100 cars

16. How many cars were sold in May?
17. Which month had the highest car sales?
18. How many more cars were sold in August than in June?
19. What was the total number of cars sold in April and June?
20. Which two months had the highest sales?

130

It is illegal to photocopy this page. Copyright © 2023, Richard W. Fisher

Final Review – Graphing on a Number Line and Graphing Ordered Pairs

Use the number line to state the coordinates of the given points.

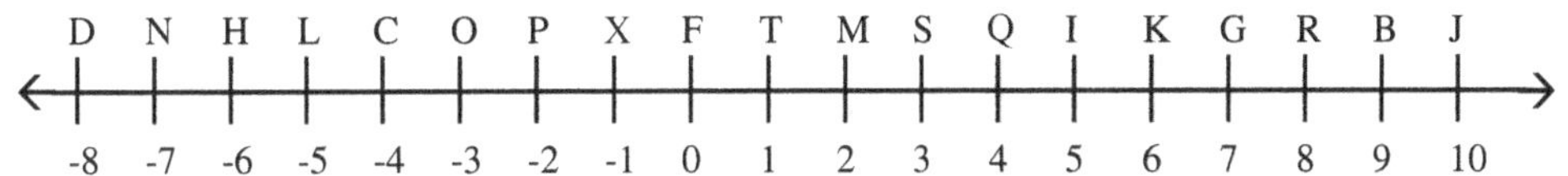

1. C   2. B, F, and J   3. S, M, and N   4. R, T, C, and D

Solve each equation and graph each solution on the number line. Be sure to label your answers. Also, place each solution in the answer column.

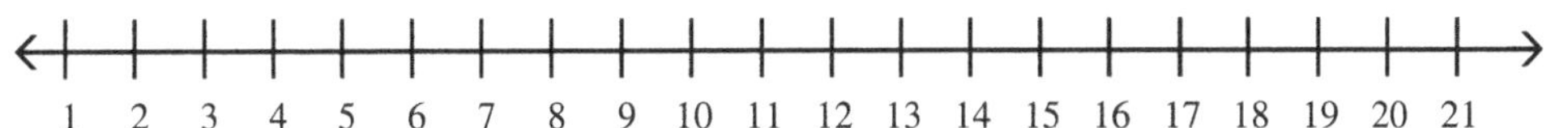

5. $n + 2 = 5$ 6. $x - 2 = 4$

7. $3y = 15$ 8. $\dfrac{m}{2} = 4$

Use the coordinate system to find the ordered pair associated with each point.

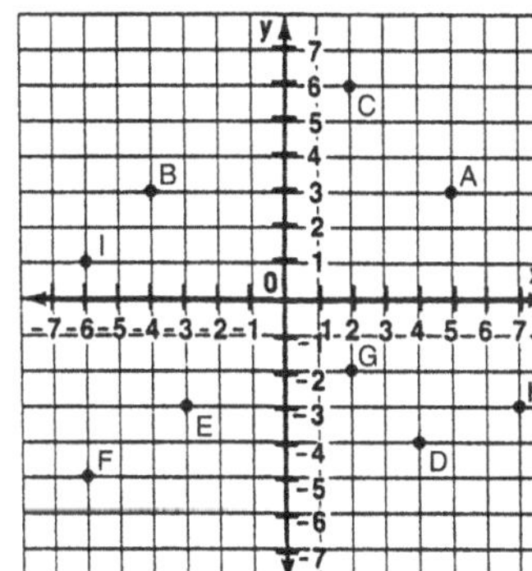

9.   A   10. I   11. D

12. F   13. C   14. E

Use the coordinate system to find the point associated with each ordered pair.

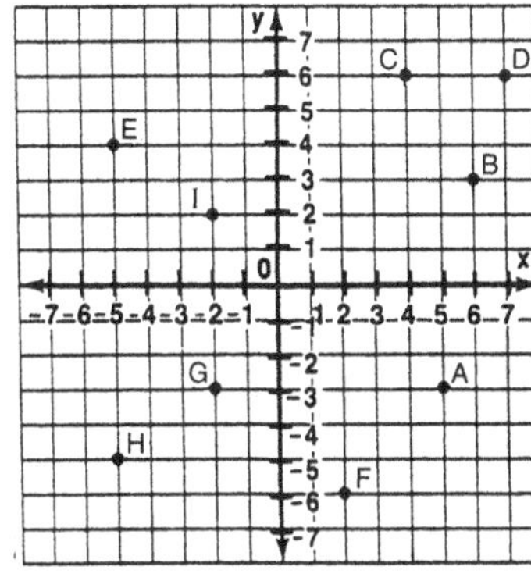

15. (6, 3)   16. (-2, 2)

17. (-5, -5)   18. (7, 6)

19. (5, -3)   20. (4, 6)

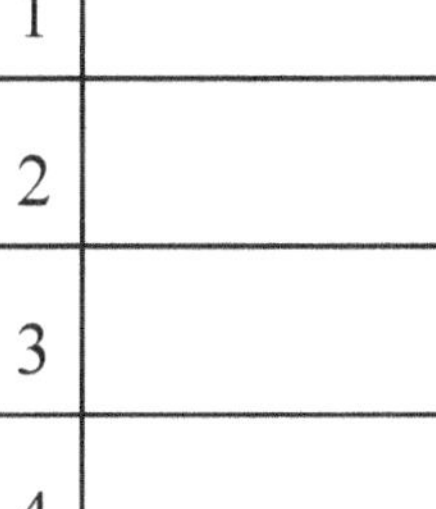

131

It is illegal to photocopy this page. Copyright © 2023, Richard W. Fisher

There are 3 red marbles, 2 blue marbles, 4 white marbles, and 1 green marble in a can. What is the probability for each of the following if a single marble is drawn from the can?

1. a red marble   2. a white marble   3. a red or white marble

4. not a blue marble 5. a white, red or blue marble   6. not a blue marble

Use the spinner to find the probability of spinning once and landing on each of the following.

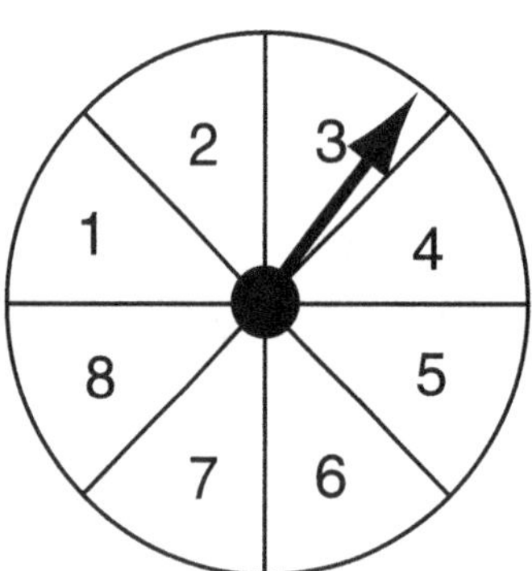

7. a four   8. an even number

9. a six   10. a one or a five

11. a number less than four

12. a one, a six or a two

Arrange the data in order from least to greatest, then answer the questions.

3, 3, 1, 8, 5

13.   What is the range?   14.   What is the mode?

15.   What is the mean?   16.   What is the median?

8, 5, 5, 5, 6, 1, 5

17.   What is the mode?   18.   What is the mean?

19.   What is the range?   20.   What is the median?

It is illegal to photocopy this page. Copyright © 2023, Richard W. Fisher

Answer Key

Page 9	Page 10	Page 11
1. 73	1. 72	1. 65
2. 737	2. 339	2. 5,528
3. 17	3. 158	3. 686
4. 10	4. 42	4. 188
S. 55	S. 579	S. 233
S. 768	S. 4,276	S. 181
1. 97	1. 5,001	1. 15
2. 95	2. 552	2. 42
3. 78	3. 6,555	3. 423
4. 407	4. 8,257	4. 683
5. 740	5. 9,527	5. 171
6. 685	6. 5,347	6. 728
7. 73	7. 2,355	7. 112
8. 250	8. 7,725	8. 511
9. 457	9. 836	9. 252
10. 30	10. 2,039	10. 52
Problem Solving: 67 students	Problem Solving: 935	Problem Solving: 53 kids

Page 12	Page 13	Page 14
1. 121	1. 33	1. 234
2. 51	2. 743	2. 64
3. 398	3. 64	3. 237
4. 335	4. 8	4. 32
S. 44	S. 954	S. 105
S. 14	S. 315	S. 2,592
1. 44	1. 78	1. 129
2. 224	2. 381	2. 150
3. 437	3. 7,653	3. 928
4. 138	4. 1,004	4. 144
5. 224	5. 64	5. 279
6. 14	6. 523	6. 2,275
7. 25	7. 47	7. 1,040
8. 284	8. 63	8. 1,360
9. 23	9. 101	9. 756
10. 272	10. 959	10. 875
Problem Solving: $8.00	Problem Solving: 32 seats	Problem Solving: 98 crayons

It is illegal to photocopy this page. Copyright © 2023, Richard W. Fisher

Page 15

1. 92
2. 846
3. 74
4. 463

S. 1,058
S. 6,132
1. 192
2. 516
3. 546
4. 2,852
5. 720
6. 7,015
7. 480
8. 3,712
9. 4,092
10. 3,960
Problem Solving: 192 desks

Page 16

1. 1,824
2. 900
3. 251
4. 691

S. 138
S. 1,296
1. 78
2. 132
3. 1,581
4. 1,410
5. 188
6. 1,058
7. 64
8. 360
9. 645
10. 9,936
Problem Solving: 1,500 sheets

Page 17

1. 21
2. 624
3. 1,027
4. 98

S. 18 r1
S. 23
1. 14 r1
2. 5 r3
3. 12 r3
4. 23 r1
5. 12 r1
6. 7 r1
7. 13 r1
8. 24 r1
9. 12 r1
10. 42
Problem Solving: 12 boxes

Page 18

1. 17 r1
2. 11 r1
3. 944
4. 47

S. 145
S. 321
1. 256
2. 312
3. 11
4. 82 r2
5. 412
6. 221
7. 121
8. 22
9. 24
10. 101
Problem Solving: 9 seats

Page 19

1. 24 r3
2. 1,491
3. 543
4. 605

S. 8
S. 71
1. 16
2. 150
3. 255
4. 34 r3
5. 16 r3
6. 6 r1
7. 212
8. 244
9. 101
10. 121
Problem Solving: $50.00 each

Page 20 Review

1. 57
2. 359
3. 637
4. 398
5. 882
6. 21
7. 505
8. 22
9. 2,918
10. 141
11. 68
12. 1,692
13. 861
14. 312
15. 4,899
16. 13 r1
17. 13 r3
18. 33 r2
19. 214
20. 336

It is illegal to photocopy this page. Copyright © 2023, Richard W. Fisher

Answer Key

Page 21

1. 54
2. 47
3. 114
4. 672

S. 1/4
S. 5/6
1. 3/8
2. 1/3
3. 2/4, 1/2
4. 5/8
5. 7/8
6. 2/3
7. 5/6
8. 1/6
9. 2/6, 1/3
10. 3/4
Problem Solving: 5 pounds

Page 22

1. 852
2. 239
3. 511
4. 214

S. 1/2
S. 1/4
1. 1/5
2. 1/3
3. 2/3
4. 2/3
5. 4/5
6. 1/3
7. 1/3
8. 1/6
9. 3/5
10. 1/2
Problem Solving: 36 crayons

Page 23

1. 3/4
2. 2/3
3. 214
4. 815

S. 1 1/2
S. 1 1/6
1. 1 3/4
2. 2 1/2
3. 1 3/5
4. 1 3/7
5. 1 1/5
6. 1 1/3
7. 2 2/5
8. 2 1/3
9. 2 1/5
10. 2 2/3
Problem Solving: 24 dancers

Page 24

1. 1 1/5
2. 1 3/7
3. 2/3
4. 1/4

S. 4/5
S. 1 1/3
1. 5/7
2. 1 2/7
3. 1 1/5
4. 3/4
5. 1
6. 1 1/8
7. 1/2
8. 4/5
9. 1
10. 2/3
Problem Solving: 1/2

Page 25

1. 3/5
2. 23
3. 1 1/5
4. 33

S. 5 1/2
S. 6 1/5
1. 5 3/5
2. 5 1/4
3. 5 1/2
4. 5 2/5
5. 6 1/7
6. 5 1/2
7. 6 1/2
8. 5 2/3
9. 5 3/5
10. 5
Problem Solving: 1/3 cup

Page 26

1. 3/4
2. 2 1/2
3. 4/5
4. 4 1/5

S. 1/4
S. 1/2
1. 1/2
2. 1/3
3. 3/7
4. 4/5
5. 3/11
6. 5/7
7. 2/5
8. 1/2
9. 1/3
10. 2/3
Problem Solving: 1/5 of a mile

It is illegal to photocopy this page. Copyright © 2023, Richard W. Fisher

Answer Key

Page 27

1. 1/2
2. 1 2/5
3. 5 5/7
4. 6 1/5

S. 3 2/5
S. 6 1/4
1. 3 3/7
2. 3 2/5
3. 6 1/3
4. 3 1/10
5. 4 7/8
6. 6 4/7
7. 3 2/9
8. 1/2
9. 3 7/10
10. 4 2/5
Problem Solving: 3 yards

Page 28

1. 1/3
2. ½
3. 5 ½
4. 71

S. 2 ½
S. 3
1. 2 1/5
2. 3 1/3
3. 3 2/3
4. 5 1/2
5. 2 2/5
6. 2 1/2
7. 3
8. 2 1/4
9. 4
10. 1 3/5
Problem Solving: 5 hours

Page 29

1. 1/5
2. 1
3. 2 2/3
4. 1/3

S. 5 1/2
S. 2 ½
1. 3/5
2. 3/4
3. 1 1/9
4. 1/4
5. 3 3/4
6. 4 1/3
7. 6 1/5
8. 3 2/3
9. 3/5
10. 1 1/2
Problem Solving:
    1 2/3 pounds

Page 30

1. 5/7
2. 1/2
3. 4 1/2
4. 2 2/3

S. 12
S. 24
1. 10
2. 8
3. 9
4. 15
5. 10
6. 16
7. 20
8. 10
9. 14
10. 30
Problem Solving: 1,200 miles

Page 31

1. 201
2. 552
3. 253
4. 21

S. 7/12
S. 3/10
1. 11/12
2. 1/6
3. 1 3/10
4. 7/9
5. 1/6
6. 1/2
7. 11/15
8. 3/4
9. 1/2
10. 1 1/6
Problem Solving:
    1 5/8 gallons

Page 32

1. 223
2. 3/4
3. 7/12
4. 1/6

S. 5 9/10
S. 8 1/10
1. 5 5/6
2. 5 7/10
3. 5 13/20
4. 4 1/12
5. 5 3/4
6. 7 5/6
7. 3 7/10
8. 5 11/15
9. 5 11/15
10. 5 9/20
Problem Solving: 216 parts

Page 33

1. 2/3
2. 2 2/5
3. 10
4. 11/15

S. 2 ¼
S. 5 ¼
1. 4/5
2. 11/12
3. 1/10
4. 2 4/5
5. 6 1/3
6. 3 3/10
7. 6 1/6
8. 1 1/2
9. 5/8
10. 4 1/2
Problem Solving: 5 dollars

136

It is illegal to photocopy this page. Copyright © 2023, Richard W. Fisher

Answer Key

Page 34

1. 12
2. 104
3. 63
4. 586

S. 6/25
S. 1 1/5
1. 1/6
2. 1/5
3. 1 1/2
4. 1 1/14
5. 2/5
6. 6/35
7. 1 1/5
8. 8/15
9. 1 1/9
10. 3/20
Problem Solving:
   2 pounds

Page 35

1. 4/15
2. 9/10
3. 0
4. 4/5

S. 3/7
S. 1 1/2
1. 3/10
2. 1/5
3. 9/16
4. 7/12
5. 1 1/7
6. 7/12
7. 9/20
8. 4/15
9. 2/3
10. 2/15
Problem Solving: 77 people

Page 36

1. 5/9
2. 2/15
3. 11/15
4. 5/12

S. 8
S. 3 1/3
1. 6
2. 4
3. 16
4. 1 1/7
5. 1 1/2
6. 6
7. 3 1/2
8. 4
9. 8
10. 2 2/3
Problem Solving: 16 girls

Page 37

1. 32 r1
2. 12
3. 3 1/3
4. 5/2

S. 1 1/4
S. 1 1/2
1. 1/2
2. 5
3. 6
4. 1
5. 3/8
6. 2
7. 2
8. 1 3/14
9. 2 2/3
10. 3/4
Problem Solving: 9 miles

Page 38

1. 4/5
2. 2/5
3. 4/7
4. 2

S. 1 1/3
S. 2/3
1. 3
2. 1 1/7
3. 3/4
4. 1/12
5. 2 1/2
6. 7
7. 1/5
8. 1 1/2
9. 8
10. 2/7
Problem Solving: $14.00

Page 39

1. 1/5
2. 1 1/2
3. 4/5
4. 1/3

S. 1 1/2
S. 1 1/4
1. 1 1/3
2. 1 4/5
3. 4/5
4. 2/15
5. 6
6. 3
7. 5
8. 1 1/12
9. 6/7
10. 6
Problem Solving: 12 yards

Page 40 Review

1. 3/5
2. 1 1/8
3. 9/10
4. 6
5. 5 5/6
6. 4 9/10
7. 1/2
8. 4 3/4
9. 2 1/6
10. 3 1/10
11. 2/15
12. 5/8
13. 6
14. 1 1/4
15. 1 2/3
16. 2/3
17. 5/6
18. 6
19. 1 1/4
20. 8/9

It is illegal to photocopy this page. Copyright © 2023, Richard W. Fisher

Answer Key

Page 41

1. 51
2. 1/6
3. 1 1/12
4. 331

S. Two and six tenths
S. Twelve and seven hundredths
1. six tenths
2. one and seven tenths
3. four and seven thousandths
4. five and sixteen hundredths
5. seventeen and twelve thousandths
6. thirteen hundredths
7. four and forty-two hundredths
8. six and three hundredths
9. six and three thousandths
10. nine hundredths

Problem Solving: 63 trading cards

Page 42

1. 552
2. 1/6
3. 3
4. 3

S. 2.6
S. 3.12
1. 9.8
2. 2.17
3. .32
4. 22.5
5. .006
6. 2.007
7. 8.2
8. 8.02
9. 2.017
10. .25

Problem Solving:
1 1/5 degrees

Page 43

1. ¾
2. 3
3. 3/8
4. 4/9

S. 7.7
S. 9.07
1. 12.32
2. .07
3. 8.9
4. 72.09
5. .16
6. 7.18
7. 6.1
8. 4.06
9. 12.6
10. 7.019

Problem Solving: 2 inches

Page 44

1. 3/5
2. 2 6/7
3. 3/10
4. 2

S. 3 2/10
S. 5 3/1000
1. 5 6/10
2. 7/100
3. 6 9/100
4. 7 9/10
5. 13 15/1000
6. 19/1000
7. 7 8/1000
8. 9 7/100
9. 8/1000
10. 5 725/1000

Problem Solving: 27 chairs

Page 45

1. 3/5
2. 3 1/2
3. Two & seven tenths
4. 3 8/100

S. >
S. >
1. >
2. <
3. <
4. <
5. <
6. <
7. >
8. >
9. <
10. >

Problem Solving: 4 days

Page 46

1. 4/5
2. 1/5
3. 1/2
4. 5 4/5

S. 8.72
S. 5.87
1. 4.54
2. 8.35
3. 7.493
4. .61
5. 13.33
6. 10.8
7. .8
8. 8.27
9. 9.26
10. 6.39

Problem Solving: 3.9 inches

Page 47

1. 8.78
2. 5.9
3. 1
4. 3.07

S. 4.08
S. 3.66
1. 3.03
2. 1.92
3. 2.67
4. 4.8
5. .08
6. 5.27
7. 1.6
8. .106
9. 6.04
10. 3.07

Problem Solving: .2 seconds

It is illegal to photocopy this page. Copyright © 2023, Richard W. Fisher

Answer Key

Page 48

1. 0
2. 3/4
3. 1
4. 1 1/3

S. 4.46
S. 5.26
1. 5.68
2. 1.8
3. 5.85
4. 3.04
5. 4.7
6. .6
7. 5.18
8. 11.04
9. 1.5
10. 5.67
Problem Solving: $4.61

Page 49

1. 72
2. 928
3. 552
4. 2,852

S. 6.86
S. 51.6
1. 6.45
2. .846
3. 212.8
4. 16.8
5. 6.96
6. 13
7. 15.42
8. 259.2
9. .936
10. 14.48
Problem Solving: 4.8 miles

Page 50

1. 6.9
2. 50.4
3. 1/10
4. 1 1/2

S. .64
S. 9.476
1. .126
2. 1.26
3. 6.42
4. 3.12
5. 1.312
6. .684
7. 1.065
8. .744
9. 14.28
10. 3.72
Problem Solving: $112.00

Page 51

1. 5
2. 3
3. 1/6
4. 10

S. 7.2
S. 21
1. 3.69
2. 13.72
3. 3.22
4. .102
5. 187.2
6. 4.86
7. .872
8. 8.64
9. 1.215
10. 57.5
Problem Solving: $21.72

Page 52

1. 21
2. 4.4
3. 2,420 r1
4. 23 r9

S. 2.3
S. 4.4
1. 1.3
2. 4.2
3. 2.11
4. .51
5. .51
6. 2.15
7. .232
8. 1.23
9. 2.23
10. 6.6
Problem Solving: 5 shelves

Page 53

1. .8
2. 3.65
3. .026
4. .006

S. .5
S. .4
1. .25
2. .6
3. .75
4. .4
5. .8
6. .2
7. .6
8. .9
9. .7
10. .3
Problem Solving: $20 dollars

Page 54

1. 2.4
2. .03
3. .115
4. .4

S. 1.2
S. .04
1. 2.3
2. .046
3. .64
4. .77
5. .5
6. 1.3
7. .8
8. .54
9. 3.1
10. 3.41
Problem Solving:
117.3 pounds

It is illegal to photocopy this page. Copyright © 2023, Richard W. Fisher

Answer Key

Page 55 Review

1. 4.47	13. 1.23
2. 4.5	14. 1.21
3. 9.2	15. .342
4. 5.11	16. .11
5. 7.33	17. .65
6. 5.3	18. 2.2
7. 6.9	19. .2
8. 9.72	20. .5
9. 7.26	
10. 84.6	
11. 4.41	
12. 9.36	

Page 56

1. 1/4
2. 4/5
3. 1/8
4. 2/3

S. 5/3
S. 9/4
1. 7/2
2. 6/5
3. 4/3
4. 5/3
5. 6/5
6. 8/3
7. 7/3
8. 6/4
9. 4/8
10. 9/3

Problem Solving: 60 miles

Page 57

1. 348
2. 59
3. 723
4. 129

S. 12%
S. 90%
1. 6%
2. 23%
3. 20%
4. 34%
5. 75%
6. 1%
7. 70%
8. 15%
9. 80%
10. 62%

Problem Solving:
   49.2 pounds

Page 58

1. 7%
2. 40%
3. 2/5
4. 5.2

S. 12%
S. 70%
1. 32%
2. 2%
3. 50%
4. 5%
5. 60%
6. 44%
7. 79%
8. 40%
9. 33%
10. 80%

Problem Solving: 60 quarts

Page 59

1. 3/4
2. 2 1/3
3. 619
4. 5 1/3

S. .12, 12/100
S. .04, 4/100
1. .16, 16/100
2. .06, 6/100
3. .75, 75/100
4. .40, 40/100
5. .01, 1/100
6. .45, 45/100
7. .12, 12/100
8. .05, 5/100
9. .50, 50/100
10. .13, 13/100

Problem Solving: 22 students

It is illegal to photocopy this page. Copyright © 2023, Richard W. Fisher

Answer Key

Page 60

1. 2.4
2. 25%
3. 7.05
4. 12.75

S. 1.2
S. 12
1. 1.56
2. 14
3. 1.2
4. 12
5. 4.8
6. 8
7. 21
8. 6
9. 16
10. .8
Problem Solving:
   3 gallons

Page 61

1. 3.12
2. .8
3. .03
4. 12

S. 3 problems
S. 30 students
1. $24
2. $800
3. 9
4. $600
5. 24
6. $16,000
7. $1,800
8. $3.50

Problem Solving: 167 miles

Page 62

1. 3
2. 1,376
3. 192
4. 381

S. 50%
S. 60%
1. 20%
2. 40%
3. 25%
4. 50%
5. 30%
6. 75%
7. 50%
8. 25%
9. 50%
10. 20%
Problem Solving: $6.00

Page 63

1. 21.36
2. 5.28
3. 6.01
4. .03

S. 40%
S. 20%
1. 80%
2. 50%
3. 80%
4. 25%
5. 50%
6. 20%
7. 20%
8. 25%
9. 50%
10. 20%
Problem Solving: $71.00

Page 64

1. 5.04
2. .4
3. 15
4. .06

S. 60%
S. 20%
1. 25%
2. 25%
3. 60%
4. 80%
5. 40%
6. 10%
7. 25%
8. 40%
9. 90%
10. 50%
Problem Solving:
   16 questions

Page 65

1. 7%
2. 90%
3. 3%
4. 70%

S. 6
S. 25%
1. 1.8
2. 7.8
3. 60%
4. 50%
5. 18
6. .8
7. 20%
8. 6
9. 15
10. 80%
Problem Solving: 24 correct

Page 66

1. 2/3
2. 1/2
3. 6/11
4. 2

S. 12
S. 25%
1. 300
2. 60%
3. $12.00
4. 75%
5. 80%
6. 4.8
7. 240
8. 60%
9. $1,200.00
10. 60%
Problem Solving: 24 miles

Page 67 Review

1. 7/2
2. 9/4
3. 8
4. 12
5. 17%
6. 70%
7. 19%
8. 60%
9. .06, 6/100
10. .15, 15/100
11. .80, 80/100
12. 2.4
13. 12
14. 75%
15. 60%
16. 20%
17. 25%
18. $8.00
19. 75%
20. 14

It is illegal to photocopy this page. Copyright © 2023, Richard W. Fisher

Answer Key

Page 68

1. 6.54
2. 1.8
3. 11.2
4. 4.6

S. answers vary
S. answers vary
1. answers vary
2. answers vary
3. answers vary
4. answers vary
5. answers vary
6. answers vary
7. answers vary
8. answers vary
9. answers vary
10. E

Problem Solving: 20 hours

Page 69

1. 1 2/5
2. 1/2
3. 1 1/2
4. 3

S. answers vary
S. answers vary
1. answers vary
2. answers vary
3. answers vary
4. answers vary
5. answers vary
6. answers vary
7. answers vary
8. answers vary
9. answers vary
10. ∠IHJ

Problem Solving: 8 pieces

Page 70

1. 75%
2. 9.6
3. 60
4. 60%

S. answers vary
S. answers vary
1. answers vary
2. answers vary
3. acute
4. obtuse
5. right
6. straight
7. answers vary
8. answers vary
9. answers vary
10. answers vary

Problem Solving: 7 meters

Page 71

1. 1 2/7
2. 5/2
3. 4/5
4. 6

S. 20 degrees
S. 110 degrees
1. 90^0
2. 160^0
3. 20^0
4. 70^0
5. 130^0
6. 50^0
7. 180^0
8. 160^0
9. 90^0
10. 130^0

Problem Solving: 40 students

Page 72

1. 80%
2. 90%
3. 2.6
4. 75%

S. 37 degrees, acute
S. 78 degrees ,acute
1. acute
2. acute
3. obtuse
4. obtuse
5. acute
6. acute
7. acute
8. acute
9. obtuse
10. acute

Problem Solving: 7 seats

Page 73

1. △DEF, acute
2. △FGH, obtuse
3. △JKL, right
4. parallel

S. parallelogram,rectangle
S. triangle
1. square; rectangle; parallelogram
2. rectangle; parallelogram
3. trapezoid
4. triangle
5. trapezoid
6. square, rectangle and parallelogram
7. parallelogram
8. rectangle; parallelogram
9. triangle
10. trapezoid

Problem Solving: $15.00

Page 74

1. 25
2. 3,100
3. 1,100
4. 288

S. scalene/right
S. equilateral/acute
1. scalene; obtuse
2. isosceles; acute
3. isosceles; right
4. scalene; acute
5. equilateral; acute
6. scalene; obtuse
7. scalene; right
8. isosceles; acute
9. equilateral; acute
10. isosceles; right

Problem Solving: 3 buses

It is illegal to photocopy this page. Copyright © 2023, Richard W. Fisher

Answer Key

Page 75

1. scalene
2. right
3. isosceles/acute
4. 24

S. 34 ft.
S. 29 ft.
1. 47 ft.
2. 48 ft.
3. 33 ft.
4. 54 ft.
5. 70 ft.
6. 41 ft.
7. 225 mi.
8. 34 ft.
9. 86 ft.
10. 34 ft.
Problem Solving: 140 ft.

Page 76

1. 42 ft.
2. 56 ft.
3. 20 %
4. 5/12

S. diameter
S. vary
1. radius
2. chord
3. answers vary
4. answers vary
5. 8 ft.
6. point P
7. answers vary
8. 16 ft.
9. answers vary
10. $\overline{XS}$, $\overline{SX}$
Problem Solving: 16 miles

Page 77

1. 254
2. 56
3. 12
4. 20%

S. 12.56 ft.
S. 25.12 ft.
1. 18.84 ft.
2. 25.12 ft.
3. 15.70 ft.
4. 6.28 ft.
5. 37.68 ft.
6. 31.4 ft.
7. 12.56 ft.
Problem Solving: 9.42 ft

Page 78

1. 9.3
2. 36.5
3. 1 1/4
4. 97.5

S. 36 sq. ft.
S. 72 sq. ft.
1. 84 sq. ft.
2. 400 sq. ft.
3. 132 sq. ft.
4. 98 sq. ft.
5. 96 sq. ft.
6. 169 sq. ft.
7. 121 sq. ft.
Problem Solving: 54 ft.

Page 79

1. 52 ft.
2. 168 sq. ft.
3. 18.84 ft.
4. 12 sq. ft.

S. P=38 ft.   A=84 sq. ft.
S. P=28 ft,   A=49 sq.ft.
1. P=48 ft.   A= 144 sq. ft.
2. P=44 ft   A=120 sq. ft.
3. P=32 ft.   A=63 sq. ft.
4. C= 9.42 ft.
5. C=12.56 ft.
6. P=32 ft.   A=39 sq. ft.
7. P=32 ft.   A=64 sq. ft.

Problem Solving: 60 ft.

Page 80

1. 734
2. 99
3. 1/3
4. 5

S. rectangular prism
   6 faces; 12 edges; 8 vertices
S. square pyramid
   5 faces; 8 edges; 5 vertices
1. cone
2. cube, 6, 12, 8
3. cylinder
4. triangular pyramid
   4, 6, 4
5. triangular prism
   5, 9, 6
6. sphere
7. 1
Problem Solving: $1.95

143

It is illegal to photocopy this page. Copyright © 2023, Richard W. Fisher

Answer Key

Page 81 Review

1. answers vary
2. answers vary
3. answers vary
4. answers vary
5. answers vary
6. answers vary
7. answers vary
8. answers vary
9. sides, equilateral, angles, acute
10. sides, scalene, angles, right
11. 29 feet
12. 15.7 feet
13. 18.84 ft.
14. 98 sq. ft.
15. 144 sq. ft.
16. 42 sq. ft.
17. 36 sq. ft.
18. square pyramid 5, 8, 5
19. 57 ft.
20. 64 ft.

Page 82

1. 677
2. 9/10
3. 10.8
4. 2.4

S. 1,2,5,10
S. 1,2,3,4,6,12
1. 1, 15, 3, 5
2. 1, 16, 2, 8, 4
3. 1, 9, 3
4. 1, 24, 2, 12, 3, 8, 4, 6
5. 1,30, 2, 15, 3, 10, 5, 6
6. 1, 18, 2, 9, 3, 6
7. 1, 8, 2, 4
8. 1, 25, 5
9. 1, 17
10. 1, 21, 3, 7
Problem Solving: $27.00

Page 83

1. 1,944
2. 1/2
3. .366
4. 20 ft.

S. 2
S. 4
1. 2
2. 3
3. 5
4. 3
5. 4
6. 4
7. 2
8. 6
9. 4
10. 4
Problem Solving: 141 points

Page 84

1. 1,20,2,10,4,5
2. 60 sq. ft.
3. acute
4. 12.56 ft.

S. 4,6,8,10
S. 0,12,18,30
1. 10, 15, 20, 25
2. 0, 6, 12, 15
3. 0, 30, 40, 50
4. 0, 4, 8
5. 22, 44
6. 24, 32, 40
7. 60, 80, 100
8. 14, 28, 35
9. 90, 120, 150
10. 27, 45
Problem Solving: 8 miles

Page 85

1. 1,2,3,4,6,12
2. 40%
3. 105 sq. ft.
4. 0,3,6,9,12

S. 12
S. 24
1. 15
2. 20
3. 30
4. 12
5. 6
6. 18
7. 10
8. 35
9. 8
10. 30
Problem Solving: $.79

Page 86

1. 11.8
2. 2.27
3. 7.2
4. .23

S. 6
S. 17
1. 4
2. 6
3. 8
4. 8
5. 11
6. 7
7. 10
8. 8
9. 14
10. 18
Problem Solving: 320 feet

Page 87

1. 12
2. 5
3. .165
4. 1.1

S. 7
S. 6
1. 7
2. 9
3. 4
4. 3
5. 14
6. 5
7. 6
8. 5
9. 25
10. 14
Problem Solving: 16 words

144

It is illegal to photocopy this page. Copyright © 2023, Richard W. Fisher

Answer Key

Page 88

1. 1 3/5
2. 2/15
3. 1/2
4. 1/8

S. 6
S. 8
1. 5
2. 12
3. 4
4. 7
5. 8
6. 7
7. 9
8. 5
9. 10
10. 11
Problem Solving: $1.76

Page 89

1. obtuse
2. 21 r1
3. 805
4. 371

S. 12
S. 8
1. 15
2. 12
3. 8
4. 3
5. 6
6. 21
7. 20
8. 5
9. 21
10. 24
Problem Solving: 90 minutes

Page 90

1. 16 ft.
2. 24 sq. ft.
3. 9.42 ft.
4. 1,2,4,5,10,20

S. 6
S. 9
1. 5
2. 9
3. 3
4. 10
5. 3
6. 2
7. 6
8. 14
9. 2
10. 1
Problem Solving: $3.60

Page 91 Review

1. 1, 20, 2, 10, 4, 5
2. 1, 24, 2, 12, 3, 8, 4, 6
3. 1, 30, 2, 15, 3, 10, 5, 6
4. 2
5. 4
6. 3
7. 6, 8, 10
8. 5, 15, 20, 25
9. 0, 7, 14, 21, 28
10. 12
11. 12
12. 9
13. 3
14. 7
15. 4
16. 8
17. 2
18. 4
19. 10
20. 4

Page 92

1. 21 ft.
2. 2/3
3. 7/12
4. 3/10

S. -5
S. -4
1. 1
2. -1
3. -9
4. 7
5. -5
6. -11
7. -11
8. 6
9. -2
10. -3
Problem Solving: 180 people

Page 93

1. 2/5, 2:5
2. n = 3
3. .044
4. 18.6

S. -1
S. 1
1. -15
2. -2
3. 8
4. -12
5. -16
6. 5
7. 10
8. -14
9. -2
10. 2
Problem Solving: 20%

Page 94

1. 32 ft.
2. 2
3. -2
4. 2

S. -7
S. 5
1. -5
2. -1
3. -8
4. 5
5. -2
6. -7
7. 1
8. -13
9. -5
10. 7
Problem Solving: 6 boxes

It is illegal to photocopy this page. Copyright © 2023, Richard W. Fisher

Answer Key

Page 95

1. -5
2. -3
3. 1
4. -5

S. -2
S. -4
1. -9
2. 2
3. -10
4. 9
5. -4
6. -16
7. 4
8. -1
9. -7
10. 1

Problem Solving: $4.52

Page 96

1. 1 ½
2. 5
3. 1 1/3
4. 4

S. 12
S. -15
1. -21
2. -35
3. 40
4. 44
5. -36
6. -28
7. 24
8. 39
9. -77
10. -306

Problem Solving: 7^0

Page 97

1. 21 r1
2. 12
3. 1
4. -10

S. -4
S. 2
1. 2
2. -5
3. -9
4. 5
5. -8
6. 9
7. -8
8. 3
9. -7
10. -3

Problem Solving: 375 dollars

Page 98 Review

1. -1
2. 1
3. -13
4. -15
5. 6
6. -9
7. -4
8. 9
9. -8
10. -1
11. 5
12. -1
13. -12
14. 15
15. -35
16. -4
17. 7
18. -8
19. -12
20. 3

Page 99

1. 12.56 ft.
2. 3.6
3. 1
4. 3

S. August
S. 80 degrees
1. November
2. July, September
3. November
4. 80°
5. October
6. 10°
7. 20°
8. 3
9. 3
10. July, September

Problem Solving: 6 dollars

Page 100

1. .4
2. 60%
3. 30%
4. 4%

S. Springdale/Winston
S. 500
1. Auberry
2. 400
3. 900
4. 200
5. Sun City
6. Mayfield
7. 2
8. 1,400
9. 500
10. 200

Problem Solving: 135 votes

Page 101

1. 25%
2. 70%
3. 90 sq. ft.
4. 2

S. 90
S. 80, 85
1. Test 1
2. 4
3. 20
4. 10
5. 3
6. 10
7. 2
8. Yes
9. 1
10. 85

Problem Solving:
102 fourth graders

Page 102

1. 1 1/5
2. ¼
3. 4 2/5
4. 3 1/3

S. Sept.
S. 400
1. May
2. July, September
3. 100
4. June, August
5. August, Sept.
6. 600
7. 300
8. July
9. 500
10. Increasing

Problem Solving: 80

It is illegal to photocopy this page. Copyright © 2023, Richard W. Fisher

Answer Key

Page 103

1. -6
2. 9
3. -24
4. 7.5

S. 23%
S. other expenses
1. 33%
2. 70%
3. Car, Clothing
4. Other expenses
5. $400
6. 35%
7. $2,000
8. answers vary
9. answers vary
10. 32%
Problem Solving: 9 miles

Page 104

1. ¾
2. 20%
3. 25%
4. 4

S. 4
S. 8 hours
1. 9 P.M.
2. 5
3. 30
4. 6 A.M.
5. 3 P.M.
6. 24
7. 6
8. Sleep, School, Play
9. 2
10. answers vary
Problem Solving: 168 sq. ft.

Page 105

1. 654
2. 1,638
3. 264
4. 27 r4

S. 5,000
S. 2,000
1. 1989
2. 9,000
3. 1989, 1991
4. 1986, 1988
5. 2,000
6. 5,000
7. 6,000
8. 10,000
9. 1987
10. 3,000
Problem Solving: $138.00

Page 106

1. 18.84 ft.
2. 52 ft.
3. 15
4. 5

S. 1990
S. 5 hours
1. 40
2. 35
3. 10
4. 1970
5. 15
6. 5 hours
7. 1970, 1990
8. 2,000
9. 5
10. 25 hours
Problem Solving: 75%

Page 107

1. 7.06
2. 4.46
3. 20.8
4. .615

S. 0
S. -7, -2, 10
1. 5, 9
2. 3, 4
3. 2, 4, 7
4. -5, -8
5. 10, 9, -6, 6
6. 9, -7, 1
7. -8, 8, 0, -3
8. 0, 7, 8
9. -6, 4, -3
10. 5, -3, 9, -8
Problem Solving: $2.30

Page 108

1. 4.2
2. 80%
3. 12
4. 75%

S. 1
S. 7
1. 3
2. 5
3. 5
4. 18
5. 4
6. 11
7. 8
8. 7
9. 4
10. 12
Problem Solving: $13.44

Page 109

1. 3
2. -13
3. -18
4. 16

S. (2, 1)
S. (-4, -2)
1. (6, 3)
2. (2, -5)
3. (-7, -3)
4. (-5, 1)
5. (4, 6)
6. (4, -3)
7. (-3, -5)
8. (-2, 2)
9. (2, 1)
10. (-6, 7)
Problem Solving: $8.50

Page 110

1. 576
2. 432
3. 3,144
4. 55 r1

S. B
S. A
1. C
2. D
3. F
4. M
5. E
6. J
7. H
8. G
9. K
10. I
Problem Solving: $175 dollars

It is illegal to photocopy this page. Copyright © 2023, Richard W. Fisher

Answer Key

Page 111 Review

1. Grant Falls
2. 525 feet
3. 75-100 feet
4. Snake Falls
5. Morton Falls
6. 13%
7. 20%
8. $600
9. 10%
10. 41%
11. 72°
12. 30°
13. June
14. August
15. 10°
16. 300
17. 200
18. 1,000
19. 900 pounds
20. Salmon, Cod, Snapper

Page 112 Review

1. -1
2. -8, 2, 9
3. -2, -4, 4
4. 1, 6, 5, -7
5. 4
6. 12
7. 9
8. 18
9. (-4, 4)
10. (6, -3)
11. (5, 6)
12. (-2, -4)
13. (-6, -3)
14. (2, -2)
15. B
16. G
17. H
18. I
19. C
20. F

Page 113

1. 1 1/6
2. 3/4
3. 1/2
4. 5
S. 3/12
S. 7/12
1. 2/12
2. 1/12
3. 9/12
4. 10/12
5. 9/12
6. 5/12
7. 10/12
8. 6/12
9. 6/12
10. 4/12
Problem Solving: $9.50

Page 114

1. 53.9
2. 3.53
3. $24.78
4. 2.05
S. 1/8
S. 4/8
1. 1/8
2. 7/8
3. 4/8
4. 4/8
5. 2/8
6. 0/8
7. 2/8
8. 5/8
9. 5/8
10. 4/8
Problem Solving: $24 dollars

Page 115

1. 30%
2. 3%
3. 80%
4. 3.25
S. R=7, M=4
S. R=6, M=6
1. Range = 6   Mode =7
2. Range = 23   Mode = 30
3. Range = 10   Mode =3,6
4. Range = 19   Mode = 9
5. Range = 7   Mode = 3
6. Range = 11   Mode = 8
7. Range = 9   Mode = 2
8. Range = 6   Mode = 91
9. Range = 9   Mode = 2
10. Range = 19 Mode = 2
Problem Solving: $18 dollars

Page 116

1. 5
2. -10
3. -24
4. 8
S. Mean = 3   Median = 3
S. Mean = 4   Median = 4
1. Mean = 3   Median = 2
2. Mean = 3   Median = 2
3. Mean = 15   Median = 15
4. Mean = 2   Median = 1
5. Mean = 6   Median = 6
6. Mean = 124 Median = 126
7. Mean = 4   Median = 4
8. Mean = 4   Median = 4
9. Mean = 5   Median = 4
10. Mean = 50 Median = 50
Problem Solving: $.38

Page 117

1. 18 ft.
2. 169 sq. ft.
3. 9.42 ft.
4. 36 sq. ft.
S. 6
S. 2
1. 5
2. 6
3. 3
4. 2
5. 8
6. 4
7. 10
8. 2
9. 5
10. 5
Problem Solving: 80

Page 118 Review

1. 3/10
2. 4/10
3. 5/10
4. 7/10
5. 8/10
6. 6/10
7. 1/8
8. 4/8
9. 5/8
10. 2/8
11. 4/8
12. 2/8
13. 8
14. 4
15. 5
16. 4
17. 2
18. 4
19. 7
20. 3

It is illegal to photocopy this page. Copyright © 2023, Richard W. Fisher

Answer Key

Page 119

1. 2 to 3, 2/3, 2:3
2. 17
3. 8
4. 15
S. 67
S. 1,075 miles
1. 548
2. $147
3. 644
4. 221
5. 520 miles
6. 60 mph
7. 104 miles
8. 46 ft.
9. 56 pieces
10. 3,573 books

Page 120

1. 1 1/5
2. 5/8
3. 1/2
4. 4
S. 3 8/15 cups
S. 5
1. 3 1/4 lbs.
2. $10
3. 4 1/2 lbs.
4. 13 minutes
5. 14 ft.
6. 3 1/4 dollars
7. 3 3/4 hours
8. 1 9/10 hours
9. 15 miles
10. 4 tires

Page 121

1. $8.63
2. $4.42
3. $13.02
4. .69
S. $7.02
S. 7.8 ft.
1. $.30
2. $185.80
3. $1.45
4. 253 miles
5. $36.25
6. $.50
7. $50.25
8. 161.1 lbs.
9. $1.78
10. $3.77

Page 122

1. 5/6
2. 6.06
3. 3/8
4. 2.07
S. 16
S. $.90
1. $1.70
2. $1.06
3. 95
4. $6
5. 11.4 inches
6. 30
7. 8 cows
8. 28 miles
9. 4 1/2 ft.
10. $13.50

Page 123 Review

1. 57
2. 730
3. 83
4. 870
5. 499
6. .57
7. 12
8. 217
9. 14
10. 468
11. 279
12. 66
13. 108
14. 2,092
15. 840
16. 820
17. 23
18. 131 r2
19. 34 r3
20. 212

Page 124 Review

1. 3/5
2. 1 1/7
3. 9/10
4. 5/6
5. 5 5/6
6. 2/5
7. 2 2/3
8. 1/4
9. 2 1/6
10. 4 1/10
11. 1/8
12. 5/8
13. 3
14. 1 1/2
15. 2 1/2
16. 3
17. 2/3
18. 1 1/3
19. 3
20. 10

Page 125 Review

1. 7.5
2. 5.42
3. 7.44
4. $5.96
5. 9.32
6. 1.9
7. 2.06
8. 5.19
9. 2.74
10. 3.7
11. 10.5
12. 3.72
13. 2.065
14. 4.83
15. 6.42
16. 10.65
17. 2.3
18. .041
19. .115
20. 1.74

Page 126 Review

1. 5/2
2. 7/6
3. 6
4. 8
5. 17%
6. 30%
7. 13%
8. 40%
9. .12, 12/100
10. .03, 3/100
11. .20, 20/100
12. 3.6
13. 1.8
14. 40%
15. 25%
16. 20%
17. 25%
18. 6 dollars
19. 25%
20. 4

It is illegal to photocopy this page. Copyright © 2023, Richard W. Fisher

Answer Key

Page 127 Review

1. answers vary
2. answers vary
3. answers vary
4. answers vary
5. answers vary
6. answers vary
7. answers vary
8. answers vary
9. sides, isosceles angles, acute
10. sides, scalene angles, obtuse
11. 51 ft.
12. 9.42 ft.
13. 6 ft.
14. 18 sq. ft.
15. 32 ft.
16. 22 ft.
17. 25 sq. ft.
18. rectangular prism, Faces 6, Edges 12, Vertices 8
19. 24 ft.
20. 64 ft.

Page 128 Review

1. 1, 12, 2, 6, 3, 4
2. 1, 15, 3, 5
3. 1, 20, 2, 10, 4, 5
4. 2
5. 4
6. 5
7. 6, 8, 10
8. 3, 9, 12
9. 0, 4, 8, 12, 16
10. 6
11. 12
12. 8
13. 2
14. 5
15. 5
16. 8
17. 3
18. 3
19. 8
20. 4

Page 129 Review

1. 2
2. -2
3. -8
4. -8
5. 2
6. -8
7. -4
8. 7
9. -5
10. -15
11. 18
12. -24
13. 32
14. -36
15. 24
16. -4
17. -4
18. 7
19. -6
20. 3

Page 130 Review

1. Ken
2. 70
3. 40
4. 200
5. Sue
6. 30%
7. 25%
8. 55%
9. Transportation
10. 15%
11. 60
12. cats
13. fish
14. 30
15. dogs
16. 400
17. August
18. 400
19. 500
20. July, August

Page 131 Review

1. -4
2. 9, 0, 10
3. 3, 2, -7
4. 8, 1, -4, -8
5. 3
6. 6
7. 5
8. 8
9. (5, 3)
10. (-6, 1)
11. (4, -4)
12. (-6, -5)
13. (2, 6)
14. (-3, -3)
15. B
16. I
17. H
18. D
19. A
20. C

Page 132 Review

1. 3/10
2. 4/10
3. 7/10
4. 8/10
5. 9/10
6. 8/10
7. 1/8
8. 4/8
9. 1/8
10. 2/8
11. 3/8
12. 3/8
13. 7
14. 3
15. 4
16. 3
17. 5
18. 5
19. 7
20. 5

150

It is illegal to photocopy this page. Copyright © 2023, Richard W. Fisher

Glossary

A

absolute value   The distance of a number from 0 on the number line. The absolute value is always positive.

acute angle   An angle with a measure of less than 90 degrees.

adjacent   Next to.

algebraic expression   A mathematical expression that contains at least one variable.

angle   Any two rays that share an endpoint will form an angle.

associative properties   For any a, b, c:
addition: $(a + b) + c = a + (b + c)$
multiplication: $(ab)c = a(bc)$

B

base   The number being multiplied. In an expression such as 4^2, 4 is the base.

C

coefficient   A number that multiplies the variable. In the term 7x, 7 is the coefficient of x.

commutative properties   For any a, b:
addition: $a + b = b + a$
multiplication: $ab = ba$

complementary angles   Two angles that have measures whose sum is 90 degrees.

congruent   Two figures having exactly the same size and shape.

coordinate plane   The plane which contains the x- and y-axes. It is divided into 4 quadrants. Also called coordinate system and coordinate grid.

coordinates   An ordered pair of numbers that identify a point on a coordinate plane.

D

data   Information that is organized for analysis.

degree   A unit that is used in measuring angles.

denominator   The bottom number of a fraction that tells the number of equal parts into which a whole is divided.

disjoint sets   Sets that have no members in common. {1,2,3} and {4,5,6} are disjoint sets.

151

It is illegal to photocopy this page. Copyright © 2023, Richard W. Fisher

distributive property   For real numbers a, b, and c: a(b + c) = ab + ac.

E

element of a set   Member of a set.

empty set   The set that has no members. Also called the null set and written Ø or { }.

equation   A mathematical sentence that contains an equal sign (=) and states that one expression is equal to another expression.

equivalent   Having the same value.

exponent   A number that indicates the number of times a given base is used as a factor. In the expression n^2, 2 is the exponent.

expression   Variables, numbers, and symbols that show a mathematical relationship.

extremes of a proportion   In the proportion $\frac{a}{b} = \frac{c}{d}$, a and d are the extremes.

F

factor   An integer that divides evenly into another.

finite   Something that is countable.

formula   A general mathematical statement or rule. Used often in algebra and geometry.

function   A set of ordered pairs that pairs each x-value with one and only one y-value. (0,2), (-1,6), (4,-2), (-3,4) is a function.

G

graph   To show points named by numbers or ordered pairs on a number line or coordinate plane. Also, a drawing to show the relationship between sets of data.

greatest common factor   The largest common factor of two or more numbers. Also written GCF. The greatest common factor of 15 and 25 is 5.

grouping symbols   Symbols that indicate the order in which mathematical operations should take place. Examples include parentheses ( ), brackets [ ], braces { }, and fraction bars —— .

H

hypotenuse   The side opposite the right angle in a right triangle.

152

It is illegal to photocopy this page. Copyright © 2023, Richard W. Fisher

I

identity properties of addition and multiplication   For any real number a:
addition: $a + 0 = 0 + a = a$
multiplication: $1 \times a = a \times 1 = a$

inequality   A mathematical sentence that states one expression is greater than or less than another. Inequality symbols are read as follows: $<$ less than
$\leq$ less than or equal to
$>$ greater than
$\geq$ greater than or equal to

infinite   Having no boundaries or limits. Uncountable.

integers   Numbers in a set. ...-3, -2, -1, 0, 1, 2, 3...

intersection of sets   If A and B are sets, then A intersection B is the set whose members are included in both sets A and B, and is written $A \cap B$. If set A = {1,2,3,4} and set B = {1,3,5}, then $A \cap B$ = {1,3}

inverse properties of addition and multiplication   For any number a:
addition: $a + -a = 0$
multiplication: $a \times 1/a = 1 \ (a \neq 0)$

inverse operations   Operations that "undo" each other. Addition and subtraction are inverse operations, and multiplication and division are inverse operations.

L

least common multiple   The least common multiple of two or more whole numbers is the smallest whole number, other than zero, that they all divide into evenly. Also written as LCM. The least common multiple of 12 and 8 is 24.

linear equation   An equation whose graph is a straight line.

M

mean   In statistics, the sum of a set of numbers divided by the number of elements in the set. Sometimes referred to as average.

means of a proportion   In the proportion $\frac{a}{b} = \frac{c}{d}$, b and c are the means.

median   In statistics, the middle number of a set of numbers when the numbers are arranged in order of least to greatest. If there are two middle numbers, find their mean.

mode   In statistics, the number that appears most frequently. Sometimes there is no mode. There may also be more than one mode.

multiple   The product of a whole number and another whole number.

153

It is illegal to photocopy this page. Copyright © 2023, Richard W. Fisher

N

natural numbers   Numbers in the set 1, 2, 3, 4,... Also called counting numbers.

negative numbers   Numbers that are less than zero.

null set   The set that has no members. Also called the empty set and written Ø or { }.

number line   A line that represents numbers as points.

numerator   The top part of a fraction.

O

obtuse angle   An angle whose measure is greater than 90° and less than 180°.

opposites   Numbers that are the same distance from zero, but are on opposite sides of zero on a number line. 4 and -4 are opposites.

order of operations   The order of steps to be used when simplifying expressions.
1. Evaluate within grouping symbols.
2. Eliminate all exponents.
3. Multiply and divide in order from left to right.
4. Add and subtract in order from left to right.

ordered pair   A pair of numbers (x,y) that represent a point on the coordinate plane. The first number is the x-coordinate and the second number is the y-coordinate.

origin   The point where the x-axis and the y-axis intersect in a coordinate plane. Written as (0,0).

outcome   One of the possible events in a probability situation.

P

parallel lines   Lines in a plane that do not intersect. They stay the same distance apart.

percent   Hundredths or per hundred. Written %.

perimeter   The distance around a figure.

perpendicular lines   Lines in the same plane that intersect at a right (90°) angle.

pi   The ratio of the circumference of a circle to its diameter. Written π. The approximate value for π is 3.14 as a decimal and $\dfrac{22}{7}$ as a fraction.

plane   A flat surface that extends infinitely in all directions.

point   An exact position in space. Points also represent numbers on a number line or coordinate plane.

It is illegal to photocopy this page. Copyright © 2023, Richard W. Fisher

positive number Any number that is greater than 0.

power An exponent.

prime number A whole number greater than 1 whose only factors are 1 and itself.

probability What chance, or how likely it is for an event to occur. It is the ratio of the ways a certain outcome can occur and the number of possible outcomes.

proportion An equation that states that two ratios are equal. $\frac{4}{8} = \frac{2}{4}$ is a proportion.

Pythagorean theorem In a right triangle, if c is the hypotenuse, and a and b are the other two legs, then $a^2 + b^2 = c^2$.

Q

quadrant One of the four regions into which the x-axis and y-axis divide a coordinate plane.

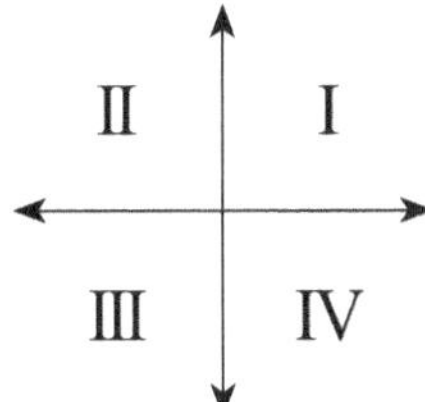

R

range The difference between the greatest number and the least number in a set of numbers.

ratio A comparison of two numbers using division. Written a:b, a to b, and a/b.

reciprocals Two numbers whose product is 1. $\frac{2}{3}$ and $\frac{3}{2}$ are reciprocals because $\frac{2}{3} \times \frac{3}{2} = 1$.

reduce To express a fraction in its lowest terms.

relation Any set of ordered pairs.

right angle An angle that has a measure of 90°.

rise The change in y going from one point to another on a coordinate plane. The vertical change.

run The change in x going from one point to another on a coordinate plane. The horizontal change.

S

scientific notation A number written as the product of a numbers between 1 and 10 and a power of ten. In scientific notation, $7{,}000 = 7 \times 10^3$.

set A well-defined collection of objects.

slope Refers to the slant of a line. It is the ratio of rise to run.

It is illegal to photocopy this page. Copyright © 2023, Richard W. Fisher

Glossary

solution   A number that can be substituted for a variable to make an equation true.

square root   Written $\sqrt{\ }$. The $\sqrt{36} = 6$ because $6 \times 6 = 36$.

statistics   Involves data that is gathered about people or things and is used for analysis.

subset   If all the members of set A are members of set B, then set A is a subset of set B. Written $A \subset B$. If set A = {1,2,3} and set B = {0,1,2,3,5,8}, set A is a subset of set B because all of the members of a set A are also members of set B.

U

union of sets   If A and B are sets, the union of set A and set B is the set whose members are included in set A, or set B, or both set A and set B. A union B is written $A \cup B$. If set = {1,2,3,4} and set B = {1,3,5,7}, then $A \cup B$ = {1,2,3,4,5,7}.

universal set   The set which contains all the other sets which are under consideration.

V

variable   A letter that represents a number.

Venn diagram   A type of diagram that shows how certain sets are related.

vertex   The point at which two lines, line segments, or rays meet to form an angle.

W

whole number   Any number in the set 0, 1, 2, 3, 4...

X

x-axis   The horizontal axis on a coordinate plane.

x-coordinate   The first number in an ordered pair. Also called the abscissa.

Y

y-axis   The vertical axis on a coordinate plane.

y-coordinate   The second number in an ordered pair. Also called the ordinate.

156

It is illegal to photocopy this page. Copyright © 2023, Richard W. Fisher

Important Symbols

<	less than		π	pi
≤	less than or equal to		{ }	set
>	greater than		| |	absolute value
≥	greater than or equal to		$.\overline{n}$	repeating decimal symbol
=	equal to		1/a	the reciprocal of a number
≠	not equal to		%	percent
≅	congruent to		(x,y)	ordered pair
( )	parenthesis		⊥	perpendicular
[ ]	brackets		| |	parallel to
{ }	braces		∠	angle
...	and so on		∈	element of
• or ×	multiply		∉	not an element of
∞	infinity		∩	intersection
a^n	the nth power of a number		∪	union
√	square root		⊂	subset of
Ø, { }	the empty set or null set		⊄	not a subset of
∴	therefore		△	triangle
°	degree			

It is illegal to photocopy this page. Copyright © 2023, Richard W. Fisher

Multiplication Table

x	2	3	4	5	6	7	8	9	10	11	12
2	4	6	8	10	12	14	16	18	20	22	24
3	6	9	12	15	18	21	24	27	30	33	36
4	8	12	16	20	24	28	32	36	40	44	48
5	10	15	20	25	30	35	40	45	50	55	60
6	12	18	24	30	36	42	48	54	60	66	72
7	14	21	28	35	42	49	56	63	70	77	84
8	16	24	32	40	48	56	64	72	80	88	96
9	18	27	36	45	54	63	72	81	90	99	108
10	20	30	40	50	60	70	80	90	100	110	120
11	22	33	44	55	66	77	88	99	110	121	132
12	24	36	48	60	72	84	96	108	120	132	144

Commonly Used Prime Numbers

2	3	5	7	11	13	17	19	23	29
31	37	41	43	47	53	59	61	67	71
73	79	83	89	97	101	103	107	109	113
127	131	137	139	149	151	157	163	167	173
179	181	191	193	197	199	211	223	227	229
233	239	241	251	257	263	269	271	277	281
283	293	307	311	313	317	331	337	347	349
353	359	367	373	379	383	389	397	401	409
419	421	431	433	439	443	449	547	461	463
467	479	487	491	499	503	509	521	523	541
547	557	563	569	571	577	587	593	599	601
607	613	617	619	631	641	643	647	653	659
661	673	677	683	691	701	709	719	727	733
739	743	751	757	761	769	773	787	797	809
811	821	823	827	829	839	853	857	859	863
877	881	883	887	907	911	919	929	937	941
947	953	967	971	977	983	991	997	1009	1013

It is illegal to photocopy this page. Copyright © 2023, Richard W. Fisher

Squares and Square Roots

No.	Square	Square Root	No.	Square	Square Root	No.	Square	Square Root
1	1	1.000	51	2,601	7.141	101	10201	10.050
2	4	1.414	52	2,704	7.211	102	10,404	10.100
3	9	1.732	53	2,809	7.280	103	10,609	10.149
4	16	2.000	54	2,916	7.348	104	10,816	10.198
5	25	2.236	55	3,025	7.416	105	11,025	10.247
6	36	2.449	56	3,136	7.483	106	11,236	10.296
7	49	2.646	57	3,249	7.550	107	11,449	10.344
8	64	2.828	58	3,364	7.616	108	11,664	10.392
9	81	3.000	59	3,481	7.681	109	11,881	10.440
10	100	3.162	60	3,600	7.746	110	12,100	10.488
11	121	3.317	61	3,721	7.810	111	12,321	10.536
12	144	3.464	62	3,844	7.874	112	12,544	10.583
13	169	3.606	63	3,969	7.937	113	12,769	10.630
14	196	3.742	64	4,096	8.000	114	12,996	10.677
15	225	3.873	65	4,225	8.062	115	13,225	10.724
16	256	4.000	66	4,356	8.124	116	13,456	10.770
17	289	4.123	67	4,489	8.185	117	13,689	10.817
18	324	4.243	68	4,624	8.246	118	13,924	10.863
19	361	4.359	69	4,761	8.307	119	14,161	10.909
20	400	4.472	70	4,900	8.367	120	14,400	10.954
21	441	4.583	71	5,041	8.426	121	14,641	11.000
22	484	4.690	72	5,184	8.485	122	14,884	11.045
23	529	4.796	73	5,329	8.544	123	15,129	11.091
24	576	4.899	74	5,476	8.602	124	15,376	11.136
25	625	5.000	75	5,625	8.660	125	15,625	11.180
26	676	5.099	76	5,776	8.718	126	15,876	11.225
27	729	5.196	77	5,929	8.775	127	16,129	11.269
28	784	5.292	78	6,084	8.832	128	16,384	11.314
29	841	5.385	79	6,241	8.888	129	16,641	11.358
30	900	5.477	80	6,400	8.944	130	16,900	11.402
31	961	5.568	81	6,561	9.000	131	17,161	11.446
32	1,024	5.657	82	6,724	9.055	132	17,424	11.489
33	1,089	5.745	83	6,889	9.110	133	17,689	11.533
34	1,156	5.831	84	7,056	9.165	134	17,956	11.576
35	1,225	5.916	85	7,225	9.220	135	18,225	11.619
36	1,296	6.000	86	7,396	9.274	136	18,496	11.662
37	1,369	6.083	87	7,569	9.327	137	18,769	11.705
38	1,444	6.164	88	7,744	9.381	138	19,044	11.747
39	1,521	6.245	89	7,921	9.434	139	19,321	11.790
40	1,600	6.325	90	8,100	9.487	140	19,600	11.832
41	1,681	6.403	91	8,281	9.539	141	19,881	11.874
42	1,764	6.481	92	8,464	9.592	142	20,164	11.916
43	1,849	6.557	93	8,649	9.644	143	20,449	11.958
44	1,936	6.633	94	8,836	9.695	144	20,736	12.000
45	2,025	6.708	95	9,025	9.747	145	21,025	12.042
46	2,116	6.782	96	9,216	9.798	146	21,316	12.083
47	2,209	6.856	97	9,409	9.849	147	21,609	12.124
48	2,304	6.928	98	9,604	9.899	148	21,904	12.166
49	2,401	7.000	99	9,801	9.950	149	22,201	12.207
50	2,500	7.071	100	10,000	10.000	150	22,500	12.247

It is illegal to photocopy this page. Copyright © 2023, Richard W. Fisher

Fraction/Decimal Equivalents

Fraction	Decimal	Fraction	Decimal
$\frac{1}{2}$	0.5	$\frac{5}{10}$	0.5
$\frac{1}{3}$	0.3	$\frac{6}{10}$	0.6
$\frac{2}{3}$	0.6	$\frac{7}{10}$	0.7
$\frac{1}{4}$	0.25	$\frac{8}{10}$	0.8
$\frac{2}{4}$	0.5	$\frac{9}{10}$	0.9
$\frac{3}{4}$	0.75	$\frac{1}{16}$	0.0625
$\frac{1}{5}$	0.2	$\frac{2}{16}$	0.125
$\frac{2}{5}$	0.4	$\frac{3}{16}$	0.1875
$\frac{3}{5}$	0.6	$\frac{4}{16}$	0.25
$\frac{4}{5}$	0.8	$\frac{5}{16}$	0.3125
$\frac{1}{8}$	0.125	$\frac{6}{16}$	0.375
$\frac{2}{8}$	0.25	$\frac{7}{16}$	0.4375
$\frac{3}{8}$	0.375	$\frac{8}{16}$	0.5
$\frac{4}{8}$	0.5	$\frac{9}{16}$	0.5625
$\frac{5}{8}$	0.625	$\frac{10}{16}$	0.625
$\frac{6}{8}$	0.75	$\frac{11}{16}$	0.6875
$\frac{7}{8}$	0.875	$\frac{12}{16}$	0.75
$\frac{1}{10}$	0.1	$\frac{13}{16}$	0.8125
$\frac{2}{10}$	0.2	$\frac{14}{16}$	0.875
$\frac{3}{10}$	0.3	$\frac{15}{16}$	0.9375
$\frac{4}{10}$	0.4		

It is illegal to photocopy this page. Copyright © 2023, Richard W. Fisher

www.ingramcontent.com/pod-product-compliance
Lightning Source LLC
Chambersburg PA
CBHW081146160726
47997CB00020B/2659

9 798988 820710